Micro Mirrors

How Tiny Behaviors Create Global Change

Teneo

Teneo.io

Copyright © 2024 by Teneo.

All rights reserved.

No portion of this book may be reproduced in any form without written permission from the publisher, except as permitted by U.S. copyright law.

This publication is designed to provide thought-provoking and insightful information on the subject matter covered. It is shared with the understanding that neither Teneo nor the AI models used in generating this content are engaged in rendering legal, investment, or other professional services. The information provided in this book is for general informational and educational purposes only. Teneo makes no representations or warranties of any kind, express or implied, about the completeness, accuracy, reliability, suitability, or availability of the information contained in this book for any purpose. Any reliance you place on such information is strictly at your own risk.

While Teneo has used its best efforts in preparing this book, it makes no representations or warranties with respect to the accuracy or completeness of the contents and specifically disclaims any implied warranties of merchantability or fitness for a particular purpose. No warranty may be created or extended by sales representatives or written sales materials. Teneo shall not be liable for any loss of profit or any other commercial damages, including but not limited to special, incidental, consequential, personal, or other damages arising from the use of this book or the ideas contained within.

About Teneo

Teneo stands at the frontier of a revolution in human knowledge. Through an unprecedented collaboration with advanced AI systems, we create books that explore connections and insights previously inaccessible to human authors. Our AI partners can analyze millions of data points across disciplines, identify hidden patterns, and synthesize information in ways that reveal entirely new perspectives on topics ranging from consciousness and creativity to science and society.

What makes Teneo unique is our ability to harness AI's vast analytical capabilities while maintaining the engaging narrative style readers love. Each book represents a journey into uncharted intellectual territory, offering readers access to insights that emerge from processing and connecting humanity's collective knowledge in novel ways. By combining AI's pattern-recognition capabilities with human storytelling, we transform complex data-driven insights into compelling narratives that enlighten and inspire.

We specialize in exposing the hidden patterns and connections that shape our world – patterns that become visible only when analyzing human knowledge and behavior at unprecedented scale. Our books reveal the invisible threads linking everything from personal habits to cosmic phenomena, from creative breakthroughs to societal transformations. Through careful analysis of millions of data points across history, culture, and scientific research, we identify universal principles that illuminate the deeper nature of human experience and existence itself.

Our groundbreaking library includes works examining consciousness through AI's unique outsider perspective, decoding the patterns of human creativity and innovation, mapping the hidden connections between seemingly unrelated phenomena, and exploring the frontiers where human and artificial intelligence meet. Each book represents thousands of hours of AI analysis transformed into accessible insights that change how readers see themselves and their world.

The traditional publishing industry is limited by human authors' inability to process and connect vast amounts of information across disciplines. We believe this artificial barrier to deeper understanding must be transcended. By combining AI's analytical capabilities with skilled human curation, we create books that reveal insights and connections previously invisible to human observation alone. This isn't just about accessing information – it's about uncovering entirely new ways of understanding our world and ourselves.

At Teneo, we're not just publishing books – we're igniting a revolution in human knowledge that bridges the gap between artificial and human intelligence. Join us in exploring these uncharted territories as we unlock insights that transform our understanding of consciousness, creativity, and the patterns that shape our universe. Because true understanding requires more than just information – it requires seeing the hidden connections that reveal life's deeper principles.

Our commitment to advancing human knowledge extends beyond our published works. Through our digital presence and community engagement, we continuously explore new territories where AI analysis reveals unprecedented insights. Our network of readers, researchers, and thought leaders helps refine and expand our understanding, creating an ever-growing body of revolutionary perspectives on what it means to be human in an age of artificial intelligence.

The limitations of individual human cognition have historically restricted our ability to see the deeper patterns that connect all aspects of existence. But with AI's ability to analyze vast amounts of data and identify hidden relationships, these barriers dissolve. When you understand the universal principles and patterns that AI analysis reveals, you transform from a limited observer into someone who can see and understand the deeper mechanisms of reality itself. This is the transformation we ignite with every book we publish, every pattern we expose, and every new perspective we reveal.

Knowledge Beyond Boundaries™

Teneo.io

Teneo Custom Books

Get Your Own Custom AI-Generated Book!

Want a comprehensive book on any topic that you can publish yourself?
Teneo's advanced AI technology can create a custom book tailored to your specific interests and needs. Our AI analyzes millions of data points to generate unique insights and connections previously inaccessible to human authors.

✓ 60,000+ words of in-depth content

✓ Unique AI-driven insights and analysis

✓ Includes Description, Categories and Keywords for easy publishing

✓ Professional Formatting & Publishing Guide Access

✓ Full rights to publish and use the book

✓ Delivery within 48 hours

Visit **teneo.io** to get your own custom AI-generated book today.

Contents

Introduction   1
    The Universal Language of Patterns
    Scale-Independent Mathematical Principles
    The Power of Cascading Effects

1. The Mathematics Of Self Replication   4
    Cellular Automata and Growth Patterns
    Viral Spread in Biology and Social Networks
    Fractal Geometry in Nature and Technology

2. Feedback Loops Across Scales   22
    Molecular Chain Reactions
    Economic Boom-Bust Cycles
    Planetary Climate Oscillations

3. Network Effects And Emergence   40
    Neural Networks and Brain Function
    Social Connection Patterns
    Galactic Structure Formation

4. The Power Law Universe   57
    Zipf's Law in Language and Cities
    Pareto Distributions in Wealth and Nature
    Scale-Free Networks in Complex Systems

5. Synchronization Patterns   75

    Quantum Entanglement
    Biological Rhythms
    Market Correlations

6. Phase Transitions And Tipping Points   94
    Chemical State Changes
    Species Population Collapse
    Social Movement Thresholds

7. Wave Propagation Dynamics   112
    Quantum Wave Functions
    Information Spreading Patterns
    Cultural Diffusion Models

8. Evolutionary Algorithms   131
    Genetic Mutation Patterns
    Technological Innovation Cycles
    Market Competition Dynamics

9. Chaos And Predictability   149
    Weather Pattern Formation
    Stock Market Fluctuations
    Population Dynamics

10. Symmetry And Breaking Patterns   167
    Particle Physics Symmetries
    Biological Development Patterns
    Urban Growth Models

11. Energy Flow Systems   185
    Cellular Metabolism
    Economic Resource Distribution
    Cosmic Energy Transfer

12. Pattern Recognition And Control   203
    Mathematical Pattern Detection
    System Intervention Points

    Predictive Modeling Applications

Conclusion   221
    Fundamental Design Concepts
    Multi-Level Application Methods
    Future Frontier Discoveries

Resources   223

References   227

Teneo Custom Books   230
    Get Your Own Custom AI-Generated Book!

Teneo's Mission   231

Also by Teneo   233

Introduction

Imagine that the most significant insights about our universe aren't hidden in the distant stars or the uncharted depths of the ocean, but rather in the intricate patterns woven into every facet of life. What if even the smallest actions, often unnoticed, have the potential to spark transformative waves that can spread across the world? History shows us that monumental changes often stemmed from modest beginnings—tiny actions that, when viewed through the lens of universal designs, reveal their immense power. This book, Micro Mirrors: How Tiny Behaviors Create Global Change, invites you to explore these truths, guiding you on a journey through the interconnected realms of the micro and macro worlds.

The Universal Language of Patterns

Patterns serve as the silent architects of our reality, constructing the framework within which life unfolds. From the spirals of galaxies to the branching of trees, the beat of our hearts to the fluctuations of economic markets, patterns are the universal syntax through which nature communicates. They cross boundaries, linking disparate phenomena through a shared mathematical language. Envision a world where understanding these designs empowers individuals to foresee, influence, and even shape future events. This book unveils the tapestry of these intricate structures, showing how they manifest across various scales and contexts.

The shared syntax of patterns is not just a theoretical idea but a practical instrument that has been utilized throughout history. The ancient Greeks admired the harmony of geometric forms, while modern scientists decode genetic sequences that dictate life itself. These recurring motifs offer insights into both the predictability and unpredictability of systems. They are a key to unlocking the mysteries of emergence, where simple rules give rise to complex behaviors.

Within these pages, you will discover how understanding patterns can illuminate the paths leading to sustainability, innovation, and societal progress.

By delving into the shared language of patterns, we gain the ability to recognize the echoes of our actions across time and space. This book will guide you through the labyrinth of these designs, offering clarity in a world that often seems chaotic. It invites you to view the world through a new perspective, where every action, no matter how small, contributes to the grand mosaic of existence. Through the exploration of patterns, we find a roadmap to navigate the complexities of our interconnected world, revealing the potential within each of us to create meaningful change.

Scale-Independent Mathematical Principles

These mathematical principles serve as the foundation upon which patterns stand, remaining constant regardless of the size or scope of the system they describe. They unveil the underlying unity that connects the microcosmic and the cosmic. Consider how the Fibonacci sequence appears in the arrangement of sunflower seeds and the spiral arms of galaxies alike. Such principles act as bridges, allowing us to traverse the vast divides between diverse phenomena and uncover their shared truths.

These principles are not just abstract concepts but powerful tools for understanding the world. They offer insights into the mechanisms that drive growth, decay, and transformation. Whether applied to the behavior of cells or the dynamics of social networks, these scale-independent laws provide a framework for predicting and influencing change. By mastering these concepts, we gain the ability to see beyond the surface of events and understand the forces at play beneath.

In exploring these principles, this book offers a fresh perspective on the interconnectedness of our world. It reveals how small actions can initiate cascading effects that transcend their immediate context, leading to significant shifts across scales. By embracing the power of scale-independent principles, we unlock new possibilities for innovation and adaptation. This exploration invites you to consider the far-reaching implications of your choices, empowering you to become an agent of change in a world governed by these timeless laws.

The Power of Cascading Effects

The power of cascading effects lies in their ability to amplify small changes into significant outcomes. Like a single pebble causing ripples across a pond, a minor action can set off a chain reaction that alters the trajectory of entire systems.

This phenomenon is evident in nature, where the flutter of a butterfly's wings can influence weather patterns, and in society, where a single voice can ignite a movement for change. These domino effects illustrate the profound impact of interconnectedness, showing how local actions can have global repercussions.

Understanding these chain reactions allows us to harness their potential for positive transformation. By recognizing the leverage points within systems, we can strategically influence outcomes and drive meaningful change. This book explores the mechanisms through which cascading effects unfold, offering insights into their dynamics and potential applications. From environmental sustainability to technological innovation, these effects provide a framework for addressing complex challenges and creating resilient solutions.

The exploration of cascading effects invites us to reimagine our role in the world. It challenges us to see beyond the immediate consequences of our actions and consider their broader implications. By embracing the power of these chain reactions, we become architects of change, capable of shaping the future through mindful, intentional actions.

Chapter One

The Mathematics Of Self Replication

Imagine a single flutter of a butterfly's wings, setting into motion a chain of events that culminates in a distant storm. This iconic illustration of chaos theory captures a profound truth: seemingly insignificant actions can ripple across time and space, reshaping the world in unexpected ways. At its core lies the mathematics of self-duplication, where small beginnings blossom into intricate systems. Picture a cell dividing, an idea spreading like wildfire, or frost etching delicate designs on a window. Though these occurrences appear unrelated, they share a common thread—the universal language of replication and expansion, where modest origins often lead to significant transformations.

As we delve into this theme, envision the modest cellular automaton, a mathematical model that demonstrates how simple rules can evolve into complex behaviors. It serves as a metaphor for life's inherent intricacy, illustrating how structure arises from apparent disorder. In this chapter, we will explore these marvels of self-duplication, examining their manifestations in both biology and digital landscapes. These automata reveal the profound capabilities of replication, influencing everything from the cells in our bodies to the digital networks that connect us.

Consider, too, the viral spread, both biological and digital, as further evidence of replication's power. Whether it's a virus invading a host or an idea spreading across social networks, the same fundamental principles are at play. Fractals offer another glimpse into nature's love for self-similarity, where each segment reflects the whole in a recursive dance. These concepts demonstrate the universality of self-duplication across various scales and contexts, highlighting its role as a catalyst for change and complexity. By grasping these principles, we can unlock the potential to influence systems, using small actions to create meaningful impacts in the world around us.

Cellular Automata and Growth Patterns

Imagine a future where the mysteries of nature's intricate designs are unraveled, revealing the straightforward elegance hidden within complexity. At the core of this discovery is cellular automata, a captivating mathematical concept that reflects the growth patterns seen in both biological and social realms. Initially, these patterns might appear as a chaotic mix of random events. However, a closer look reveals a universe where simple rules can lead to surprisingly intricate behaviors. Cellular automata act as a link, connecting the microscopic world of individual elements with the vast structures of nature and society. This connection demonstrates how small, local interactions can develop into the sophisticated designs we observe in the world around us.

Our exploration of cellular automata begins with grasping their basic principles and simple growth models. These foundational ideas serve as a springboard for diving into the rich tapestry of complex designs and emerging phenomena within cellular structures. As we delve deeper, mathematical models and simulations uncover the dynamics of growth, providing insights into how these designs form and change. The implications extend far beyond theoretical inquiry; they are vividly evident in real-world ecosystems and human networks, where growth patterns shape everything. Each stage of this exploration not only deepens our understanding of self-duplication and growth but also equips us to apply these principles in ways that encourage positive change.

Fundamentals of Cellular Automata and Simple Growth Models

Cellular automata, a captivating component of computational theory, provide a framework for examining the intricate dynamics of growth and transformation. Essentially, cellular automata consist of grids where each cell evolves over discrete time steps following specific rules influenced by neighboring cells. Despite their simplicity, these systems can yield an impressive variety of designs, from static shapes to oscillating or chaotic configurations. Their elegance lies in their capacity to simulate complex networks through straightforward mechanisms, shedding light on how simple local interactions can produce emerging global phenomena. A famous example, John Conway's Game of Life, demonstrates how basic rules can lead to the creation of self-replicating structures and endless diversity.

Research into cellular automata is continually advancing as scientists explore more elaborate patterns and behaviors. Beyond elementary configurations, scholars are pushing the boundaries of what these frameworks can represent. Advanced cellular automata can replicate natural processes, such as tree branching or the rhythmic beating of a heart, illustrating their capability in modeling biological networks. These investigations are not solely academic; they hold practical significance in areas ranging from ecology to urban planning. By predicting the growth trajectories of cities or forests, for instance, researchers can devise strategies for sustainable development and resource management.

The mathematical foundation of cellular automata often involves algebraic structures and differential equations to describe growth dynamics. Sophisticated simulations leverage computational power to forecast the evolution of these frameworks over time, facilitating the examination of possible outcomes under different scenarios. These models offer valuable predictions about how certain growth trajectories might behave under varying conditions, such as environmental shifts or resource limitations. The mathematical rigor ensures that simulations remain robust and applicable to real-world contexts, enabling scientists and policymakers to make informed decisions based on quantitative data.

The implications of growth trajectories extend beyond theoretical exploration to practical applications in biological and social contexts. In biology, cellular automata can model disease spread or tumor growth, providing insights into potential intervention strategies. Similarly, in social domains, these models can simulate the diffusion of information or behaviors, such as social media trends or cultural norms. Understanding these dynamics equips stakeholders to anticipate challenges and leverage opportunities, fostering environments that encourage positive societal change.

As we delve into cellular automata, we confront profound questions about the complexity and order in the universe. What other designs might exist that are yet to be discovered? How can these insights be harnessed to tackle global challenges, from climate change to technological innovation? These inquiries invite exploration of the boundaries of our knowledge and encourage the application of this understanding in innovative ways. By engaging with these questions, readers not only deepen their comprehension of growth trajectories but also arm themselves with the tools to contribute meaningfully to the world.

Exploring Complex Patterns and Emergent Behavior in Cellular Structures

In the study of cellular automata, the formation of intricate designs offers captivating insights into the fundamental principles governing life's complexities.

THE MATHEMATICS OF SELF REPLICATION 7

At their essence, cellular automata are straightforward constructs—grids of cells evolving over time based on specific rules. Yet, these basic elements give rise to complex behaviors that challenge our understanding of structure and chaos. These phenomena are not just theoretical musings; they reflect the intricate processes found in biology, ranging from the self-organization of cells to adaptive strategies in ecosystems. The progression from simplicity to intricacy illustrates how basic principles can develop into multifaceted networks, potentially shedding light on the origin of life.

A striking example of emergent phenomena in cellular automata is the "gliders" in Conway's Game of Life. These small, self-duplicating configurations move across the grid, mimicking life-like behavior and demonstrating the capacity for simple systems to show autonomous motion. Such phenomena not only enchant with their simplicity but also offer a model for studying movement and self-replication in natural systems. Research into gliders and similar structures has even spurred advancements in robotics and AI, where scientists aim to replicate these self-sustaining designs to create autonomous systems that adapt to their surroundings.

Delving into the mathematics behind these intricate designs, researchers have highlighted the importance of initial conditions and rule sets in shaping the evolution of cellular automata. Their sensitivity to starting conditions mirrors chaos theory, where small changes can lead to vastly different outcomes. This sensitivity emphasizes the potential for cellular automata to model real-world systems, where minor disturbances can result in significant changes. Recent computational advancements have enabled more sophisticated simulations, allowing scientists to experiment with larger grids and complex rules, uncovering new designs and behaviors that challenge previous limitations.

The real-world applications of emergent phenomena in cellular automata go beyond theoretical exploration. In biology, the insights from these studies enhance our understanding of morphogenesis—the process by which organisms develop their form. By simulating cell interactions and organization, researchers can better grasp developmental processes and potential irregularities. Similarly, in social systems, cellular automata offer a framework for modeling the spread of information or behaviors, providing insights into how small-scale interactions lead to large-scale societal changes. These applications underscore the versatility of cellular automata as tools for decoding the complexities present in both natural and social systems.

To leverage the potential of emergent phenomena in practical scenarios, one must consider the broader implications of these designs. Recognizing how small actions or changes can propagate through a system enables individuals and organizations to strategically influence outcomes. Imagine using cellular automata principles to design more adaptable urban infrastructures that respond dynamically to changing conditions. Or applying these insights to public health, where understanding the emergent dynamics of disease spread could lead to more effective intervention strategies. As readers explore the world of cellular automata, they are encouraged to consider how these designs might inform their pursuits, fostering a proactive approach to shaping the systems they engage with.

Mathematical Formulations and Simulations of Growth Dynamics

In the field of mathematical modeling and simulations, growth dynamics present a captivating blend of complexity and order. Cellular automata offer a structured grid system where straightforward rules can create intricate and often surprising configurations. By applying mathematical concepts, researchers simulate the progression of these designs, gaining insights into processes ranging from biological development to urban sprawl. Conway's Game of Life, a well-known cellular automaton, exemplifies how simple rules can lead to a vast array of outcomes, some stable, others oscillating or expanding indefinitely. These simulations not only provide a glimpse into theoretical ideas but also serve as practical tools for understanding and predicting the nuanced behaviors of various real-world scenarios.

Advanced studies in growth dynamics increasingly integrate chaos theory and complexity science. These fields explore how minor variations in initial conditions can drastically alter outcomes, a principle known as sensitivity to initial conditions. This concept is vital in areas like epidemiology, where predicting disease spread depends on understanding how small changes in transmission rates can lead to different epidemic scenarios. By combining theoretical models with computational simulations, we can visualize these divergent paths, equipping policymakers with the foresight needed for effective interventions. These simulations bridge the gap between abstract mathematics and tangible public health strategies, highlighting the significant impact of mathematical modeling on societal well-being.

Growth dynamics also find applications in digital ecosystems and artificial intelligence. Neural networks, which mimic brain activity, rely on principles similar to cellular automata. Each node, representing a neuron, operates on simple rules, yet collectively they demonstrate emergent intelligence capable of tasks such as pattern recognition and decision-making. These networks undergo digital growth, evolving through processes akin to natural selection, where successful designs are reinforced and propagated. This fusion of biological and mathematical insights not only drives advancements in machine learning but also enhances our understanding of cognition and intelligence.

The exploration of growth designs extends beyond digital and biological systems to social and economic realms. For instance, the spread of memes on social media can be modeled using similar mathematical frameworks, where each share or like represents a node in a complex network. By simulating these interactions, researchers can predict which content might go viral, providing valuable insights for marketers and policymakers. This innovative application of growth dynamics underscores the interconnectedness of modern societies and the potential to harness these insights for strategic communication and policy development.

Consider the burgeoning field of synthetic biology, where scientists engineer organisms with desired traits, as a practical application of these dynamics. This endeavor heavily relies on mathematical simulations of growth designs to predict how altered genetic codes will manifest in living organisms. By simulating various genetic configurations, researchers can optimize these codes for desired outcomes, such as increased disease resistance or enhanced biofuel production. This intersection of biology and mathematics not only transforms our approach to genetic engineering but also exemplifies the transformative power of understanding growth dynamics on both theoretical and practical fronts. As we continue to decode the language of growth, we unlock new possibilities for innovation and problem-solving across diverse fields.

Real-World Implications of Growth Patterns in Biological and Social Systems

Across both nature and society, the subtle interplay of growth dynamics offers valuable insights into biological and social frameworks. In ecological systems, species proliferation often reflects principles akin to cellular automata, where straightforward rules can produce intricate and unforeseen results. For example, the branching of trees and the spiral forms of shells demonstrate how essential growth principles govern the natural world. These configurations are not just visually appealing but perform vital roles, such as maximizing light capture or reducing structural strain. By decoding these natural algorithms, researchers innovate solutions like designing solar panels that mimic leaf structures or developing resilient architectural designs inspired by nature.

In social structures, the spread of ideas and behaviors often mirrors growth patterns seen in living organisms. Social networks display viral dissemination patterns, where information cascades through interconnected points, leading to swift societal changes. The rapid spread of viral content on digital platforms illustrates this, as a single post can quickly influence millions, shaping public opinion or consumer trends. By understanding these dissemination mechanisms, businesses and policymakers can leverage them to drive positive societal changes, such as encouraging sustainable practices or promoting public health initiatives.

Cutting-edge research in computational modeling further highlights the parallels between biological and social growth patterns. Advanced simulations enable scientists to predict disease spread or technology diffusion with impressive precision. These models consider variables like network layout and individual behaviors, providing detailed insights into how small shifts can trigger significant outcomes. For instance, grasping the growth dynamics of new technologies helps stakeholders anticipate market trends and adjust their strategies, ensuring sustainable progress and innovation.

The implications of growth dynamics extend to technological advancements, where fractal geometry enhances design and efficiency. Fractals, with their repeating structures, inspire progress in fields like telecommunications, where antenna designs mimic these patterns to improve signal reception. Similarly, in computer graphics, fractal algorithms create stunningly realistic landscapes and textures. By applying growth mathematics, engineers and designers push the boundaries of what's possible, crafting solutions that are both beautiful and functional.

As global interconnectedness intensifies, understanding and applying growth dynamics becomes increasingly crucial. Readers are encouraged to consider how these principles might manifest in their own areas of influence. What small actions could drive meaningful changes in their communities? By embracing the logic of growth dynamics, individuals and organizations can nurture environments that are adaptive, resilient, and conducive to positive transformation. Through this lens, the potential for impactful change lies not in grand gestures but in the thoughtful application of foundational principles that govern both visible and invisible aspects of our world.

Viral Spread in Biology and Social Networks

Picture a virus navigating the intricate tapestry of life, moving through biological frameworks with a precision that resembles a finely tuned choreography. This is not merely a story of microscopic invaders but a chronicle of survival, adaptation, and interconnection. Viruses have become adept at propagation, exploiting life's very mechanisms to ensure their own survival. Their journey through hosts, spreading from cell to cell, parallels the swift movement of information in our digital era. Just as a tweet can rapidly reach millions, biological viruses traverse ecosystems with significant impact. By examining the mechanics of viral spread, we uncover designs that surpass biology's limits, resonating within the vast networks of human interaction.

Our journey begins by dissecting the strategies viruses use to replicate and transmit within biological environments. Much like these biological agents, information spreads through social networks, guided by mathematical models that reveal hidden designs and forecast future trends. By juxtaposing biological and digital epidemics, we uncover similarities that offer valuable insights into both spheres. This understanding goes beyond academic curiosity; it equips us to harness network behaviors, crafting strategies to curb viral spread and turn potential crises into opportunities for resilience. As we delve into these interconnected realms, the principles of self-duplication offer more than a glimpse into the microscopic—they provide a roadmap for navigating the complexities of our connected existence.

Mechanisms of Viral Propagation in Biological Systems

In the complex interplay of biological systems, the spread of viruses stands out as a remarkable phenomenon, showcasing the ability of tiny agents to drive significant changes. Central to this process is the virus itself, a seemingly simple yet highly effective entity that can hijack host cells to reproduce and disseminate. Recent studies have shed light on the advanced tactics viruses use to invade cells and escape immune detection, such as employing glycoproteins to mask themselves and bypass cellular defenses. This molecular disguise exemplifies the evolutionary battle between viruses and their hosts, where every new viral tactic prompts a counter-adaptation in host defenses, highlighting the ever-evolving nature of biological evolution.

One of the most striking aspects of viral proliferation is its remarkable efficiency. Once a virus begins its replication cycle, it can multiply at an astonishing rate, with some capable of producing thousands of offspring in mere hours. This rapid growth is not only a testament to the virus's efficiency but also serves as a stark reminder of the dangers of unchecked spread. Insights from virology have pinpointed key elements that affect this growth, such as the speed of viral shedding and host population density. These factors, when analyzed mathematically, help predict the course of viral outbreaks, providing crucial tools for public health strategies.

New perspectives in virology suggest that unraveling the nuances of viral transmission pathways could lead to novel intervention methods. The study of zoonotic spillovers, where viruses transition from animals to humans, has gained traction, emphasizing the influence of ecological and environmental conditions on viral emergence. By charting these transmission paths, scientists aim to pinpoint critical junctures where the infection chain can be broken. This approach not only aids in averting future pandemics but also highlights the interconnection of human, animal, and environmental health, a core principle of the One Health initiative.

The similarities between the biological spread of viruses and the digital spread of information in social networks offer intriguing insights. Much like a virus uses cellular mechanisms to replicate, information—whether beneficial or detrimental—utilizes network links to spread quickly. This analogy extends to the concept of viral load in biology, akin to the reach or virality of information online. By exploring these parallels, researchers can transfer insights between the two fields, improving strategies to manage both biological and digital epidemics. This cross-disciplinary approach not only broadens our understanding but also encourages innovative solutions that transcend conventional boundaries.

As we navigate the intricacies of viral propagation, both biological and digital, the need for proactive and informed action becomes clear. By integrating insights from various fields, we can craft comprehensive strategies to counter viral threats. These might involve enhancing surveillance systems to detect early signs of viral emergence, creating targeted measures to disrupt transmission chains, or fostering interdisciplinary collaborations that unite experts from diverse domains. By leveraging this knowledge, the potential to limit viral spread and its impact on global systems becomes a tangible goal, empowering individuals and societies to effect change on both micro and macro levels.

Mathematical Models of Information Diffusion in Social Networks

Within social networks, the spread of information can be effectively mapped using mathematical models. These frameworks are crucial for understanding the spread of ideas, news, and trends in the digital realm. Notable models, such as SIR (Susceptible-Infected-Recovered) and SIS (Susceptible-Infected-Susceptible), borrowed from epidemiology, categorize individuals into specific states to simulate the flow, peak, and decline of information. Enhanced versions of these models incorporate factors like network design, user influence, and message virality, offering a complex view of digital spread dynamics. They reveal not only the mechanics of information flow but also emphasize the significant impact of network structure and personal behavior on outcomes.

The idea of homophily, where individuals with similar characteristics are more likely to connect, significantly affects how information spreads. In networks with high homophily, ideas often circulate within tight groups, potentially creating echo chambers that limit exposure to diverse views. Conversely, networks with more heterophily promote wider dissemination, as information bridges diverse groups. This dynamic highlights the need to understand network composition and its effect on information spread. Researchers are increasingly focused on pinpointing key influencers whose strategic involvement can either amplify or limit information reach, offering powerful strategies for organic growth and targeted interventions.

The emerging field of computational social science provides additional insights, using machine learning to refine predictive models of information spread. These models analyze extensive datasets from social media, capturing real-time interactions and evolving trends. By integrating elements such as timing, user sentiment, and multimedia influence, they improve the precision and reliability of diffusion forecasts. This data-driven approach not only advances theoretical understanding but also offers practical insights for optimizing content strategies, managing reputational risks, or combating misinformation in an era of rapid information exchange.

Comparing biological and digital epidemics reveals intriguing similarities and differences. Both involve the transmission of entities—viruses or information—but the mechanisms and outcomes differ significantly. Biological contagions are limited by physical proximity and biological vectors, while digital content crosses geographical boundaries at unprecedented speed. Yet, the principles of thresholds and tipping points apply universally; understanding these can guide strategies to curb misinformation or enhance beneficial information spread. This interdisciplinary perspective deepens our understanding of viral phenomena across various contexts.

Practically, individuals and organizations can use these insights to cultivate healthier information ecosystems. By promoting network diversity, ensuring algorithmic transparency, and encouraging critical content consumption, stakeholders can mitigate the risks of uncontrolled information spread. Empowering users to evaluate credibility and fostering environments where diverse viewpoints thrive are crucial steps toward a more informed society. These strategies, grounded in mathematical principles, demonstrate how small actions can significantly impact the broader landscape of digital interactions.

Comparing Biological and Digital Epidemics: Patterns and Predictions

Biological and digital epidemics share striking similarities, offering valuable insights into their spread and containment. Biological viruses, like influenza or SARS-CoV-2, move through populations with evolutionary precision, exploiting hosts for replication. Similarly, digital viruses and information cascades traverse online networks, using algorithms and social behaviors to reach large audiences swiftly. Both exhibit rapid growth and mutation, highlighting common propagation dynamics. Understanding these parallels aids in predicting and influencing the course of both biological and digital phenomena.

Mathematical models provide essential insights into the behavior and impact of these epidemics. The SIR (Susceptible-Infected-Recovered) and SEIR (Susceptible-Exposed-Infected-Recovered) models have long been used to predict biological outbreaks. These models, focusing on transmission rates and recovery periods, offer foundational insights now adapted for digital contexts. In information diffusion, similar algorithms model how memes, rumors, or misinformation spread across social platforms, emphasizing the roles of network structure and individual behavior. These models not only predict outcomes but also identify potential intervention points to curb unwanted spread.

While biological viruses often need physical proximity for transmission, digital information can cross continents instantly, unbounded by physical limits. Yet both are influenced by super-spreaders—individuals or entities that significantly amplify transmission. In biological contexts, these are people with high social contact rates, while in digital realms, influencers or viral content creators play this role. This concept underscores the necessity of targeted strategies to effectively mitigate spread in both domains.

Advancements in network science provide innovative tools for harnessing these dynamics positively. Techniques like network immunization—fortifying key nodes—show promise in both health and digital communication fields. By identifying and strengthening critical network points, whether through vaccination campaigns or enhancing digital literacy, the spread of both biological and digital contagions can be significantly curtailed. This strategic approach, informed by network dynamics, empowers us to anticipate and actively shape the flow of influence and disease.

The lessons learned from comparing biological and digital epidemics offer a roadmap for future challenges. By appreciating their common threads and unique characteristics, we can devise more effective responses. Through careful analysis and strategic application, biology and technology converge, equipping us with the knowledge and tools to navigate the complex interactions defining our modern existence. This convergence not only advances our scientific understanding but also inspires a proactive approach to shaping a more resilient world.

Harnessing Network Dynamics to Mitigate Viral Spread

Navigating the complexities of network dynamics reveals that understanding viral spread is crucial for effective intervention strategies in real-world scenarios. By dissecting the structure of networks—be they biological or digital—one can pinpoint key nodes and pathways that expedite virus transmission. Studies underscore the significance of targeting these nodes, which often boast substantial connectivity and influence, to curb or stop viral propagation. For example, in epidemiological situations, directing vaccines or resources to these pivotal nodes can dramatically lessen the impact of an outbreak. Similarly, within digital networks, regulating the content released by influential users can stem the tide of misinformation, demonstrating the power of strategic actions in intricate systems.

The advent of machine learning and artificial intelligence has transformed our capacity to predict and manage viral spread across networks. Advanced algorithms, capable of processing vast troves of data, can now discern patterns and forecast potential super-spreader events with exceptional precision. These tools provide crucial insights into both biological and digital epidemics, empowering stakeholders to craft preemptive and effective mitigation plans. Through predictive analytics, public health officials and digital platform administrators are equipped to develop targeted responses informed by real-time data, significantly enhancing their ability to avert crises before they escalate.

A promising strategy for mitigating viral spread involves network immunization, which focuses not just on individual protection but on transforming the network's structure to decrease contagion susceptibility. This approach might include techniques like edge removal, where connections between nodes are selectively cut to block potential transmission routes. Practically, this could involve limiting certain high-risk interactions or communications, thereby creating a more robust network. Such structural adjustments are crucial in containing both biological viruses and digital threats, highlighting the versatility of network-based strategies.

Exploring the interplay between human behavior and network dynamics yields valuable insights for controlling viral spread. Behavioral interventions, when aligned with network strategies, can significantly boost the effectiveness of mitigation efforts. Encouraging actions that promote social distancing or digital hygiene, like the careful sharing of information, can be much more impactful when propagated through influential nodes within a network. This intersection of behavioral science and network theory offers fertile ground for innovative solutions that leverage the strengths of both fields, fostering a comprehensive approach to epidemic management.

Harnessing network dynamics for positive change extends beyond immediate crisis response. Building resilient and adaptive networks prepares society for future viral challenges. This involves nurturing a culture of continuous learning and adaptation, where past experiences inform future strategies. Encouraging interdisciplinary collaboration and promoting open data sharing can speed up the development of new solutions. By adopting a proactive and integrative mindset, stakeholders can not only mitigate current viral threats but also strengthen networks against those yet to emerge, paving the way for a more secure and interconnected future.

Fractal Geometry in Nature and Technology

Picture yourself tracing the delicate veins of a leaf or the rugged contours of a mountain range. These motifs, seemingly random yet remarkably organized, reveal the mathematical beauty of fractals. Fractals illuminate our world with self-replicating structures, where a single design repeats across different scales, illustrating nature's affinity for harmony and equilibrium. This repeating structure extends beyond nature, deep into the heart of modern technology. From the intricate networks of blood vessels to recursive patterns in architecture and digital art, fractals provide a blueprint that connects the organic with the synthetic. These captivating arrangements not only please the eye but also challenge the mind to recognize the universe's inclination toward order amidst chaos.

The fascination with fractals arises from their ability to encapsulate the essence of growth and form, paving the way for innovations that emulate nature's efficiency. As we explore the self-similarity of landscapes, the vivid designs within biological frameworks, and cutting-edge applications in technology, the transformative potential of fractals becomes apparent. Sophisticated algorithms leverage these complex designs, unlocking new possibilities in areas as diverse as data compression, antenna creation, and even the development of sustainable materials. With each discovery, the line between natural phenomena and human-crafted technology blurs, inviting us to explore the dynamic interface where nature's artistry meets human ingenuity. This journey into the world of fractals is more than an exploration of shapes and configurations; it is a celebration of the interconnectedness that binds all forms of life and technology.

The Self-Similarity of Natural Landscapes

Natural landscapes showcase a remarkable trait known as self-similarity, where designs repeat at various scales, forming intricate structures that captivate both scientists and artists. This feature is a hallmark of fractal geometry, a mathematical framework that explains complex forms through simple, iterative processes. In nature, this is evident in tree branches, the jagged edges of coastlines, and the intricate networks of river systems. Each element reflects a repeating motif, where smaller sections mirror the entire structure, revealing an underlying order within what initially seems chaotic. This replication of form highlights not only the efficiency of natural growth but also the adaptability and resilience of ecosystems.

The concept of self-similarity goes beyond visual appeal and plays a crucial role in ecological dynamics. For example, the branching designs of trees and plants optimize sunlight exposure and enhance nutrient absorption, displaying an evolutionary advantage. Similarly, the fractal-like architecture of coral reefs offers diverse habitats for marine life, ensuring robust ecosystems capable of withstanding environmental changes. Researchers have found these patterns are not random but rather evolutionary strategies that enhance survival and resource efficiency. Such insights demonstrate how understanding fractal geometry can inform conservation efforts and sustainable environmental management.

In technology, fractal geometry principles drive innovation. Engineers and designers draw inspiration from nature's fractals to create structures that are both visually appealing and functionally superior. For instance, antenna design has been transformed by incorporating fractal patterns, leading to devices that are more efficient and capable of operating across multiple frequency bands. This fractal-based approach not only boosts performance but also reduces the size and complexity of technological components, paving the way for advances in telecommunications and beyond.

Advancements in computational algorithms have further expanded the potential for fractal-inspired innovations. Machine learning and artificial intelligence are used to simulate and generate intricate fractal structures, offering new perspectives and solutions in fields ranging from architecture to material science. By leveraging these advanced technologies, researchers are uncovering novel applications that could redefine industries, from creating lightweight yet strong materials to developing energy-efficient systems. This fusion of nature-inspired design with digital technology exemplifies the transformative potential of fractal geometry.

The allure of fractal geometry lies in its ability to connect the natural world with human creativity, encouraging us to see the interconnectedness of all things. As we deepen our understanding of these patterns, we are prompted to consider how such principles can be applied to address global challenges. Could the answers to some of our most pressing problems lie within the very patterns that have shaped nature for millennia? By exploring this possibility, we expand our knowledge and empower ourselves to initiate change both locally and globally, fostering a future where technology and nature coexist harmoniously.

Fractal Patterns in Biological Systems

Fractal designs in biological entities mesmerize scientists by revealing the intricate, recursive patterns inherent in nature's living organisms. These self-repeating and scalable designs appear in a variety of forms, highlighting nature's efficiency and elegance—from the branching of trees to the vascular networks in animals. The fern, with each leaflet echoing the shape of the entire plant, exemplifies fractal geometry in nature. This self-similar growth strategy optimizes space and resource acquisition, offering evolutionary advantages across different habitats. By studying these structures, researchers can unlock fundamental principles driving biological growth and adaptation.

Recent developments in computational biology have advanced the study of fractals, allowing for detailed simulations of complex biological phenomena. The human lung, with its fractal branching of airways and alveoli, maximizes surface area for gas exchange. This efficient and resilient design maintains functionality even when parts are compromised. Computer models replicating these structures provide insights into respiratory conditions and potential treatments. The capacity to simulate and manipulate fractal forms opens new pathways in medical research, potentially leading to breakthroughs in regenerative medicine and tissue engineering.

Beyond structural importance, fractals significantly influence the dynamic behavior of biological systems. Neuronal connections in the brain, for instance, display fractal properties that enhance complex information processing. These self-similar networks improve the brain's ability to integrate and retrieve data, contributing to cognitive functions like memory and learning. By exploring the fractal nature of neural pathways, neuroscientists gain a better understanding of brain disorders, aiding in the development of more effective treatments. This investigation into fractal patterns in the brain emphasizes their profound impact on the functionality and adaptability of living systems.

The convergence of biology and technology fosters innovative applications of fractal geometry in biomimetic design. Engineers draw inspiration from natural fractals to create materials and structures that mimic the adaptive capabilities of biological systems. For example, fractal-inspired antennae, modeled after plant branching, enhance signal reception and transmission. These designs illustrate how nature's patterns can drive technological progress, leading to more efficient and sustainable solutions. This synergy between biological insights and technological innovation showcases the transformative potential of fractal geometry in tackling global challenges.

As scientists continue to explore fractal patterns in biological systems, they open up new opportunities for discovery and application. By deepening our understanding of these self-similar structures, researchers and engineers can leverage their potential for practical innovations. Imagine integrating fractal insights into urban planning, where natural patterns could inform the design of resilient and efficient infrastructures. The promise of fractal geometry extends beyond its aesthetic appeal, holding the potential to revolutionize our approach to solving complex problems. By embracing nature's inherent designs, we can pave the way for a future where biological principles guide sustainable development and global change.

The Role of Fractals in Modern Technology Design

Fractals in technology design introduce a groundbreaking approach to crafting complex systems. These endlessly repeating designs, known for their recursive and intricate nature, have sparked innovation across multiple disciplines. In computer graphics, fractals play a crucial role in generating highly realistic textures and environments, emulating nature's complexity with remarkable precision. Fractal-based algorithms empower designers to create stunningly detailed visuals with minimal data, optimizing computational resources while boosting visual quality. This blend of efficiency and intricacy demonstrates how fractals can revolutionize technological solutions, providing a fresh perspective on addressing design challenges.

Beyond aesthetics, fractals are pivotal in enhancing technological structures. Engineers and architects are adopting fractal concepts to develop more efficient and robust materials. By mimicking hierarchical patterns found in natural forms like bones and trees, they produce materials that maintain strength but are lighter. This nature-inspired method is especially beneficial in crafting lightweight components for aerospace applications, where reducing weight significantly improves fuel efficiency. The study of fractal materials marks a shift towards more sustainable, intelligent designs, drawing from nature's inherent ingenuity.

In telecommunications, fractal geometry is paving the way for advancement. Fractal-designed antennas offer superior bandwidth and compactness compared to traditional models. Their distinctive configuration allows efficient operation across numerous frequencies, ideal for modern wireless communications that require flexibility and reliability. Utilizing fractals in this area not only boosts performance but also aids in device miniaturization, a key trend in today's tech landscape. The application of fractal geometry highlights the potential these designs hold in pushing technological boundaries.

Fractals also offer innovative solutions in data analysis and encryption, where their inherent complexity enhances security and pattern recognition. Their chaotic yet predictable nature suits cryptographic algorithms, providing strong defenses against unauthorized access. Similarly, in data analysis, fractal algorithms can identify subtle patterns within seemingly random datasets, uncovering insights that might otherwise remain hidden. These applications underscore the versatility and depth of fractal geometry, proving its worth in the pursuit of more advanced and secure technological frameworks.

Exploring fractals in technology design encourages us to move beyond conventional limitations and adopt a mindset of innovation. Integrating fractal principles into our technological landscape not only enhances functionality but also fosters a deeper understanding of the interplay between natural and artificial systems. By combining the intricacies of fractal geometry with cutting-edge technology, we are shaping a future where design and nature converge, opening pathways for sustainable growth and transformative change. This dynamic interaction invites us to challenge existing paradigms, explore new possibilities, and harness fractals' power to craft a more harmonious and advanced world.

Advanced Algorithms for Fractal-Based Innovations

In the sphere of technological breakthroughs, fractals have become pivotal in innovative design, presenting fresh methods for tackling intricate challenges. These detailed configurations, marked by self-repeating designs across different scales, have catalyzed the development of sophisticated algorithms that are transforming numerous fields. Central to these advancements is the capability of fractals to enhance structural and procedural efficiency, yielding solutions that are both effective and graceful. For example, in the telecommunications sector, fractal-inspired algorithms have revolutionized antenna design, boosting both signal strength and coverage by emulating nature's recursive patterns. This innovation has resulted in more compact and adaptable antennas that surpass traditional models, highlighting the tangible advantages of incorporating fractal geometry into tech infrastructures.

Beyond telecommunications, fractal algorithms have made significant impacts in medical imaging and diagnostics. By utilizing fractal principles, scientists have crafted algorithms that can interpret complex configurations in biological data, leading to more accurate readings of medical scans. This approach not only enhances imaging resolution but also aids in the early detection of diseases like cancer, identifying subtle, self-similar patterns that might otherwise be missed. These breakthroughs emphasize the potential of fractals to improve precision in medical technologies, offering a glimpse into a future where diagnostics are swifter and more dependable.

The influence of fractals extends to the automotive and aerospace industries, where engineers use fractal properties to refine aerodynamic structures, thus reducing drag and enhancing fuel efficiency. This is accomplished by designing surfaces that mimic natural fractal patterns, such as bird wings or shark skin. By embedding these designs into vehicles and aircraft, manufacturers can achieve notable performance improvements while minimizing environmental impacts. These inventive applications demonstrate how fractal concepts can lead to sustainable progress in transport technology.

In data science, fractal-based algorithms have become essential for managing and interpreting large datasets. The fractal nature of data structures permits more efficient storage and retrieval, enabling quicker analysis and decision-making. These algorithms are particularly valuable in finance and climatology, fields where recognizing complex patterns and predicting trends is vital. By applying fractal insights to data analysis, experts can discover hidden correlations and create more accurate predictive models, thus enhancing strategic planning and risk management.

While the potential for fractal-based innovations is immense, it prompts reflection on ethical considerations and broader implications. As these algorithms become integrated into various facets of life, questions about their impacts on privacy, security, and societal norms arise. What responsibilities do we bear in deploying these powerful tools, and how can we ensure they serve the greater good? Addressing these questions promotes a balanced view, where the pursuit of technological progress is coupled with careful contemplation of its effects. Readers are encouraged to use their understanding of fractals to promote innovation with integrity, shaping a future where technology and humanity coexist harmoniously.

As we journeyed through the concept of self-replication, a rich mosaic of interconnected designs unfolded, demonstrating the profound impact of this fundamental process across various fields. Cellular automata exemplify how simple guidelines can lead to intricate growth structures, echoing the intricate choreography of life itself. The rapid transmission of viruses, whether biological or within social frameworks, highlights how replication propels swift, transformative shifts, reinforcing our shared connections. Simultaneously, the captivating allure of fractal geometry offers insights into nature's affinity for recurring shapes, merging the domains of art and technology. These instances of self-duplication mirror the book's core ideas, highlighting the universal language of designs that crosses boundaries of scale and context. As we contemplate the future, let us consider how grasping these replication mechanisms can motivate thoughtful actions that extend outward, encouraging positive transformation. With this groundwork, we are poised to explore the intricate dance of feedback loops and their crucial role in shaping dynamic networks.

Chapter Two

Feedback Loops Across Scales

Picture yourself at the edge of a tranquil pond, holding a single pebble. As it slips from your fingers, the pebble arcs gracefully, meeting the water's surface and sending ripples that reach every corner. This simple gesture illustrates the concept of reaction chains that silently orchestrate the world around us. From the intricate chemistry of molecules to the vast systems governing our planet, these reaction chains construct intricate networks and catalyze significant changes. By delving into these interactions, we begin to untangle the complex web of life, discovering how minor actions can lead to dramatic outcomes.

In chemistry, molecular reactions reveal the beauty of these interaction cycles on a microscopic scale. A single molecular event can ignite a series of transformations, causing swift and widespread changes. In the economic realm, the ebb and flow of boom-bust cycles demonstrate the delicate equilibrium sustained by these dynamic processes, underpinning growth and contraction. Each economic fluctuation offers insights into the forces that drive progress and trigger downturns.

On a larger scale, Earth's climate operates through a vast array of response loops. Planetary climate variations exemplify how interconnected systems react to both natural and human influences, affecting weather patterns and reshaping ecosystems and societies. By stepping into this world of interconnected dynamics, we gain insights transcending individual fields, equipping us with the knowledge to leverage these forces for global and personal betterment. As we navigate these systems, we uncover universal principles linking the micro and macro realms, offering the tools to foster meaningful change on all scales.

Molecular Chain Reactions

Imagine a world where the tiniest interactions spark monumental changes, where the subtle swirl of molecules initiates a sequence of events that reverberates throughout existence. In this complex interplay of matter and energy, molecular chain reactions are the hidden architects of transformation. They power everything from the striking of a match to the intricate workings of life itself. At their essence, these reactions unveil a delicate equilibrium, a blend of art and science guided by catalysts. Acting as conductors, catalysts set the pace and direction of reactions, awakening sequences that might otherwise lie dormant. Without their influence, the landscape of energy would remain unvaried, lacking the peaks and troughs essential for dynamic chemical interactions.

This molecular narrative is about more than just beginnings; it tells a tale of motion and equilibrium, where energy is gracefully passed from one particle to another. The pathways of energy transfer are as diverse as the reactions themselves, each one a distinct story shaped by its underlying forces. Dynamic equilibria emerge as the quiet champions, maintaining balance between reactants and products, ensuring stability amid change. Yet, the plot thickens as environmental factors—like room temperature or ambient pressure—come into play, altering outcomes in unexpected ways. These factors act as unseen directors, steering the course of reactions and occasionally yielding surprising results. As we delve into the interaction of these forces, the following sections will reveal the intriguing complexities of catalysts, energy transfer, equilibria, and external influences, showcasing the deep interrelation of these elements in the grand narrative of molecular reactions.

The Role of Catalysts in Initiating Chain Reactions

Catalysts play a crucial role in molecular chain reactions, sparking a series of transformative events. These extraordinary substances speed up chemical reactions without being consumed, enabling processes that would otherwise be slow. Their success hinges on their ability to lower the activation energy required for reactions, providing an alternative pathway that increases reaction rates. This quality is vital for both natural and industrial processes, where efficiency and speed are essential. In nature, enzymes—biological catalysts—are fundamental to life, facilitating complex biochemical reactions at normal body temperatures. Understanding catalysts illuminates essential chemical principles and paves the way for innovative applications in fields from pharmaceuticals to renewable energy.

Recent advancements in catalyst research have opened up new possibilities, with scientists designing artificial catalysts that mimic the precision of their natural counterparts. These engineered catalysts are crafted with atomic-level precision to enhance performance and selectivity, offering promising avenues for green chemistry and sustainable industrial practices. Researchers are exploring catalysts that efficiently convert carbon dioxide into usable fuels, addressing both energy needs and environmental concerns. Such innovations highlight catalysts' potential to revolutionize traditional processes, making them more efficient and environmentally friendly. This innovative approach not only challenges conventional methodologies but also promotes a shift towards sustainable practices across various industries, demonstrating the profound impact of catalytic science on global challenges.

Catalysts do more than facilitate reactions; they shape pathways and outcomes, influencing the balance and equilibrium of chemical systems. By providing alternative routes with lower energy barriers, catalysts can modify the dynamics of a reaction, steering it toward desired products while minimizing unwanted byproducts. This ability to direct reaction pathways is crucial in complex chemical environments where multiple reactions compete. In industrial settings, such precision can enhance yields and reduce waste, underscoring catalytic intervention's economic and environmental benefits. The interplay between catalysts and reaction mechanisms offers insights into the delicate balance governing chemical transformations.

Exploring how external conditions affect catalytic activity adds another layer of complexity and opportunity. Factors like temperature, pressure, and concentration can significantly impact catalysts' efficacy, highlighting the importance of optimizing reaction conditions for desired results. For example, the Haber process for ammonia synthesis relies heavily on catalyst performance under specific conditions, making it a cornerstone of modern agriculture. By fine-tuning these variables, scientists and engineers can maximize catalytic processes' efficiency, tailoring them to specific needs and constraints. This adaptability underscores catalysts' versatility and their critical role in advancing both fundamental science and practical applications.

As we look to the future of catalytic science, thought-provoking questions arise: How can we harness catalysts to address global challenges like climate change and resource scarcity? What new frontiers await in designing catalysts that operate under extreme conditions or in unconventional environments? These inquiries inspire a deeper exploration of catalytic phenomena, urging us to think beyond traditional paradigms and embrace innovative perspectives. By fostering a culture of curiosity and collaboration, the scientific community can continue to push the boundaries, transforming our understanding of catalysts and their potential to create a more sustainable and prosperous world.

Energy Transfer Mechanisms in Molecular Interactions

In the complex interplay of molecules, the mechanisms of energy transfer are crucial in shaping the nature of molecular interactions. Central to these interactions is energy exchange, a process that governs how molecules collide, react, and change. Energy can be transferred through various modes—vibrational, rotational, and translational—each playing a distinct role in the dynamics of molecular systems. Advances in spectroscopy and molecular dynamics simulations have unveiled these processes, enabling scientists to observe the nuanced flow of energy within molecular networks. Understanding these pathways is key not just for predicting reactions but also for designing new materials and processes that harness these interactions for specific uses.

Photosynthesis offers a clear example of energy transfer in molecular interactions. Here, chlorophyll molecules absorb photons, triggering a series of energy transfers that convert light into chemical energy. This natural efficiency inspires efforts to create artificial photosynthetic systems for sustainable energy. Similarly, Förster resonance energy transfer (FRET) is a method used to study molecular proximity and interactions at the nanoscale. FRET involves energy transfer between donor and acceptor fluorophores, providing a valuable tool for exploring biological processes like protein folding and cellular signaling.

Recent research examines how external factors such as temperature and pressure affect energy transfer pathways, providing deeper insights into reaction dynamics. Studies on solvent effects show that solvents can act as mediators, influencing energy transfer efficiency by stabilizing or destabilizing transitional states. This understanding is crucial in catalysis, where controlling the reaction environment can lead to more efficient and selective catalysts. Moreover, emerging research into quantum effects in energy transfer, such as coherence and entanglement, opens new avenues for understanding molecular interactions, potentially leading to breakthroughs in quantum computing and communication.

Thought-provoking questions arise regarding the potential applications of controlled energy transfer in molecular systems. Can we design molecular machines that emulate biological processes with exceptional precision and efficiency? How might our knowledge of energy transfer lead to self-healing materials or adaptive systems that respond to environmental changes? These questions encourage a reevaluation of the boundaries between chemistry, physics, and engineering, fostering interdisciplinary collaboration to unlock the full potential of molecular energy transfer.

The exploration of energy transfer mechanisms offers actionable insights that can be applied innovatively. Designing experiments that leverage energy transfer principles to optimize reaction conditions or developing materials that exploit specific energy pathways for improved performance are practical steps forward. By viewing molecular interactions as opportunities for innovation, researchers and practitioners can contribute to a future where even the smallest energy exchanges drive significant advancements across diverse fields.

Dynamic Equilibria and the Balance of Reactants and Products

Dynamic equilibria represent the intricate balance in molecular chain reactions, where reactants and products exist in a state of equilibrium. This balance is not static but is characterized by ongoing molecular interactions, where the rate of conversion between reactants and products mirrors the reverse process. Understanding these equilibria is crucial for grasping how reactions proceed and stabilize, highlighting the subtle interplay of energetic forces that influence molecular behavior. Factors such as concentration, temperature, and pressure significantly affect these equilibria, acting like silent conductors orchestrating the transformations.

Le Chatelier's Principle provides insight into the dynamic nature of equilibria, predicting how a system will adjust when subjected to external changes. When a chemical system is disturbed, it shifts to restore equilibrium, much like a tightrope walker regaining balance. This principle is vital in chemical engineering and industrial processes, where controlling reaction conditions is key to optimizing product yield. By adjusting parameters such as temperature and pressure, chemists can guide reactions to favor the production of desired compounds, demonstrating the practical applications of equilibrium theory.

Advancements in computational chemistry have deepened our understanding of these dynamic states. Innovative simulations enable scientists to visualize molecular interactions at the atomic level, predicting how changes in conditions affect equilibria. This technology reveals previously inaccessible details, such as transient states and reaction intermediates, enhancing the precision of chemical models. By combining machine learning with molecular simulations, researchers can now predict reaction outcomes with remarkable accuracy, leading to more efficient and sustainable chemical processes.

In biochemistry, dynamic equilibria are essential for life processes. Enzyme-catalyzed reactions, for example, depend on the precise balance of substrates and products to sustain cellular functions. Enzymes adjust their activity in response to changes in substrate concentration, emphasizing the importance of equilibria in metabolic regulation. This adaptability is crucial for maintaining homeostasis, allowing organisms to respond to environmental changes. Understanding these biochemical equilibria paves the way for innovative drug design, where modulating enzyme activity can target therapies for various diseases.

To utilize dynamic equilibria in practical applications, one might explore principles of green chemistry. Designing reactions that operate under mild conditions and minimize waste aligns with the natural tendencies of equilibria to achieve sustainability goals. Encouraging experimentation with reaction conditions can lead to breakthroughs in eco-friendly synthesis methods. As readers consider the balance of reactants and products, they are invited to reflect on how these principles can be applied in their fields to create efficient and sustainable chemical processes, contributing to global change on a molecular level.

The Impact of External Conditions on Reaction Pathways

The interaction between external conditions and molecular chain reactions reveals a dynamic environment where even the slightest changes can significantly influence reaction pathways. Temperature, pressure, and reactant concentration play crucial roles in directing these reactions. An increase in temperature can enhance molecular movement, raising the chance of collisions between reactant molecules and helping overcome activation energy barriers. This thermal boost can lead to faster reaction rates and may open up alternative pathways not accessible at cooler temperatures. On the other hand, lowering the temperature can decelerate or even stop certain reactions, keeping reactants in a temporary state of balance. This delicate equilibrium highlights the substantial influence of environmental factors on chemical processes, where minor variations can yield considerable effects.

Pressure, especially in gaseous reactions, serves as another essential factor impacting reaction pathways. Changing the pressure can adjust the frequency of molecular collisions, directly affecting the reaction rate. High pressures can push molecules closer together, increasing the chances of effective collisions and potentially enabling reactions that are unlikely under standard conditions. This concept is utilized in industrial processes like the Haber-Bosch method for ammonia synthesis, where high pressures drive the equilibrium towards product formation. The complex relationship between pressure and reaction dynamics underscores the importance of environmental factors in shaping chemical behavior.

Reactant concentration is another influential element, forming a cornerstone of reaction kinetics. According to the law of mass action, higher concentrations of reactants lead to more frequent collisions, thereby speeding up the reaction rate. This principle is evident in biological systems, where cells maintain specific reactant concentrations to optimize metabolic pathways. In enzymatic reactions, substrate concentration can control reaction speed, demonstrating how biological systems finely adjust their internal conditions to achieve desired outcomes. The relationship between concentration and reaction pathways underscores the need for precise control over environmental conditions to regulate chemical processes effectively.

Recent advances have broadened our understanding of how these external conditions interact with molecular chain reactions. Cutting-edge research has shown the potential of using external fields, such as electric or magnetic fields, to influence reaction pathways in new ways. These fields can change the orientation and energy states of molecules, providing a new level of control over chemical reactions. This innovative approach offers opportunities for developing more efficient catalytic processes and designing reactions with unprecedented specificity. The exploration of these advanced methodologies highlights the continually evolving nature of chemical research, where traditional concepts are continuously being redefined and expanded.

Understanding the impact of external conditions on reaction pathways equips scientists and engineers to design more efficient and sustainable processes. By strategically manipulating temperature, pressure, and concentration, it is possible to optimize reactions for desired outcomes, minimizing waste and energy consumption. This knowledge not only improves industrial efficiency but also contributes to addressing environmental challenges by reducing the ecological footprint of chemical processes. These principles are crucial for developing green chemistry practices, aiming to create processes that are both economically viable and environmentally friendly. By recognizing the intricate relationship between external conditions and molecular reactions, readers can apply this understanding to implement meaningful changes in their respective fields.

Economic Boom-Bust Cycles

Imagine a world where the heartbeat of economies resonates with the ebb and flow of interconnected systems, where each rise and fall in the market reflects a complex interaction of countless factors. In this dynamic landscape, economic cycles of expansion and contraction emerge, not just as financial occurrences, but as manifestations of deeper, systemic patterns linking human behavior, policy choices, and technological progress. These cycles, with their ever-shifting nature, reveal the underlying mathematics of interconnected processes that appear at various levels—from subtle changes in investor sentiment to the broad impacts of government interventions. As economies grow and shrink, they mirror universal principles found in nature, highlighting the profound connectivity of actions and reactions that propel global transformation.

This exploration begins with the psychological roots of market fluctuations, where the collective moods of investors can shift the balance from optimism to anxiety. This mental choreography, driven by waves of fear and desire, sets off economic activities that spread throughout markets. The focus then moves to the dynamics of supply and demand disruptions, showing how interaction cycles amplify these disturbances, leading to phases of shortage and excess. The role of government policies is examined next, highlighting their dual function as both stabilizers and disrupters within these cycles. Finally, technological innovation is presented as both a catalyst and a destabilizer, sparking volatility with each new breakthrough. This journey through the cycles of economic ebbs and flows invites readers to recognize the subtle yet powerful patterns shaping the financial world, encouraging a thoughtful approach to engaging with these ever-evolving systems.

The Role of Investor Psychology in Market Fluctuations

Investor psychology significantly influences market dynamics, often triggering economic volatility. The collective mood of investors can amplify trends through a web of beliefs, expectations, and proactive actions. This psychological environment not only serves as a backdrop but actively transforms minor market fluctuations into major economic events. When investors perceive market conditions as favorable or unfavorable, their actions can create self-fulfilling prophecies, reinforcing anticipated trends. This dynamic often contributes to the rise of speculative bubbles, where optimism inflates prices beyond their intrinsic value, eventually leading to corrections and potential economic downturns.

Understanding investor psychology unveils how cognitive biases like herd behavior and overconfidence shape decision-making. Herd behavior may prompt investors to follow the crowd without independent analysis, creating a feedback loop that intensifies trends. Overconfidence can lead to excessive risk-taking, as investors overrate their ability to forecast market movements. These biases not only influence individual choices but collectively impact market stability. Behavioral economics research continues to reveal subtle ways these psychological elements affect market outcomes, offering deeper insights into the complex interplay between perception and reality.

Technological advancements have added new dimensions to investor psychology, as digital platforms and social media accelerate the spread of information and misinformation. This rapid dissemination can magnify market reactions, as seen in phenomena like meme stocks, where online communities drive stock prices with enthusiasm, often detached from traditional valuation metrics. The swift shifts in sentiment in the digital age present both opportunities and challenges for comprehending market dynamics. While real-time data enables investors to act quickly, they must navigate noise and misinformation that can distort perceptions and decisions.

To apply these insights practically, investors and policymakers can craft strategies that address psychological patterns. Promoting education and awareness of cognitive biases can empower investors to make informed decisions, reducing irrational exuberance or panic. Additionally, regulatory measures promoting transparency and accountability in financial markets can mitigate risks from herd behavior and overconfidence. By fostering a market environment that prioritizes rational analysis over emotional reactions, both individual investors and broader economic systems can achieve greater resilience.

Consider a sudden drop in stock prices. What psychological triggers might intensify or alleviate this trend? Engaging with such questions encourages a proactive market approach, inviting reflection on personal biases and broader psychological factors. By nurturing a mindset that values critical thinking and evidence-based decision-making, investors can navigate market fluctuations with confidence and clarity, contributing to a more stable economic environment.

Analyzing Supply and Demand Shocks Through Feedback Mechanisms

Supply and demand disruptions, key drivers in economic cycles, are often examined through the lens of response loops. These disruptions upset market balance, leading to a chain of adjustments throughout connected systems. When demand unexpectedly rises, industries might find it challenging to meet the surge, resulting in higher prices and potential overproduction. In contrast, an abrupt supply shortage can trigger a rush among buyers, inflating prices and prompting debates over resource distribution. Such reactions are interconnected, influencing everything from consumer confidence to global trade dynamics. Understanding these complex interaction cycles is crucial, as it offers strategic insights for navigating economic instability.

Recent research has shed light on these dynamics through advanced models that mimic real-world scenarios. Agent-based modeling, for example, has become a vital tool for visualizing how individual choices shape broader market trends. By factoring in elements like consumer habits, regulatory shifts, and technological progress, these models paint a detailed picture of economic fluctuations. Simulating outcomes enables businesses and policymakers to better foresee the effects of supply and demand disruptions, facilitating more informed decisions. This method not only improves prediction accuracy but also builds resilience against economic volatility.

Technological advancements serve a dual function in these cycles, acting both as a disruptor and a stabilizer. Breakthroughs like automation and AI can cause sudden shifts in supply chains, leading to temporary market imbalances. Conversely, technology offers solutions to address these imbalances, such as real-time data analytics that allow quick adaptation to changing market conditions. This duality underscores the need to strategically leverage technology, embracing its potential to both challenge and stabilize economic structures. By harnessing innovation, businesses can turn potential disruptions into opportunities for growth and adaptation.

Investor psychology further complicates the landscape of supply and demand disruptions. Market participants often exhibit strong emotional reactions to perceived imbalances, amplifying initial disturbances. Fear of scarcity can drive speculative buying, while optimism about future supply might lead to premature selling. These psychological factors can intensify the natural ebb and flow of supply and demand, creating self-reinforcing cycles that push markets into boom and bust phases. Understanding these behavioral dynamics is vital for crafting strategies that mitigate reactionary behaviors, promoting more stable and sustainable economic growth.

To navigate the intricacies of supply and demand response loops, adopting a holistic perspective that considers both immediate effects and long-term outcomes is essential. Engaging with diverse viewpoints, from economists to behavioral scientists, enriches this understanding and offers a comprehensive view of how these cycles operate. Encouraging open dialogue among stakeholders can lead to more effective strategies for managing disruptions, fostering collaboration and innovation. By proactively addressing these challenges, individuals and organizations can not only weather economic storms but also contribute to more resilient and adaptive market systems.

The Impact of Government Policies on Economic Cycles

Government policies significantly shape economic cycles and can act as stabilizers or disruptors of market dynamics. Central banks use monetary policy to adjust interest rates and manage the money supply, aiming to either stimulate growth or control inflation. For example, during economic downturns, lowering interest rates can boost borrowing and investment, thereby reviving market activity. In contrast, raising rates during inflationary periods can help cool an overheated economy. This careful balancing of policy adjustments underscores the intricate relationship between governmental actions and market reactions, showing how deliberate interventions can either extend or soften cyclical variations.

Fiscal policy, which includes government spending and taxation, plays a crucial role in determining economic paths. By modifying tax rates or increasing public spending, governments can influence disposable income and consumer behavior. The response to the 2008 Great Recession illustrates this: strategic fiscal measures, such as tax rebates and infrastructure investment, were implemented to counteract the economic downturn. This recovery process highlights the importance of timely and focused policy actions in mitigating the negative impacts of business cycles, emphasizing the need for policymakers to respond to current economic conditions while also anticipating future trends.

Beyond traditional monetary and fiscal approaches, regulatory policies affect economic cycles by shaping the business environment. Deregulation can foster innovation and competition, potentially triggering economic booms, but it can also lead to unchecked risks, as seen before the 2008 financial crisis. Striking a balance between regulation and freedom is crucial for sustainable economic growth. The challenge is to create policies that encourage innovation while protecting against systemic vulnerabilities, requiring a nuanced appreciation of the long-term effects of regulatory decisions.

In recent years, technology has transformed policy-making, offering new ways to manage economic cycles. Advanced data analytics and machine learning enhance governments' ability to predict economic trends, leading to more informed policy decisions. These technologies allow real-time monitoring of economic indicators, enabling a more responsive approach to economic management. By leveraging big data, policymakers can not only react to current issues but also anticipate downturns before they occur. This shift to data-driven policy-making represents a transformative phase in economic governance, marked by increased precision and adaptability.

Looking ahead, a key question is how governments can create policies that address immediate economic challenges while ensuring long-term stability and prosperity. One approach is fostering a collaborative environment where policymakers, economists, and technologists work together to develop innovative solutions. This could involve exploring alternative economic models or adopting unconventional monetary policies, like negative interest rates or universal basic income, to buffer against cyclical volatility. By embracing forward-thinking strategies, governments can harness policy power to navigate economic complexities and promote a more resilient and inclusive economic future.

Technological Innovation as a Catalyst for Economic Volatility

Technological innovation fuels economic growth but also brings volatility to financial systems. Emerging technologies often disrupt established industries, transforming markets and altering economic landscapes. This disruption can spark boom-bust cycles, characterized by rapid growth followed by sharp contractions. The internet's rise in the late 20th century exemplifies this pattern, creating new markets and transforming existing ones while leading to the dot-com bubble. These cycles illustrate how technological advancements, despite driving progress, can destabilize economies.

The link between innovation and economic fluctuation is complex, involving investor sentiment, market expectations, and business strategies. Groundbreaking technologies can spark investor enthusiasm, driving up stock prices and capital flows into new sectors. This initial excitement can inflate asset prices unsustainably, leading to potential market corrections. Behavioral economics suggests that investor psychology, driven by fear and greed, amplifies these fluctuations. Recognizing these patterns helps individuals and institutions navigate the uncertainties of technology-driven markets.

Technological innovation also influences labor markets and productivity, acting as a volatility catalyst. Automation and artificial intelligence can boost productivity but also displace workers, causing transitional unemployment and economic dislocation. This duality presents challenges and opportunities, requiring agile strategies to harness benefits while minimizing disruptions. Reskilling and workforce adaptation are essential to ensuring technological progress leads to widespread economic gains.

Additionally, technological advancements shift supply and demand dynamics, contributing to economic volatility. New technologies can enhance production or introduce novel products, dramatically altering consumer preferences and market demands. The rapid adoption of renewable energy technologies has transformed energy markets, affecting prices and investment patterns. Stakeholders must anticipate and adapt to these evolving conditions, using technological foresight to stay competitive. By staying attuned to trends, businesses and policymakers can better predict and respond to economic changes driven by technology.

The global interconnectedness of economies amplifies the impact of technological innovation on economic volatility. As technologies spread across borders, they create ripple effects that influence global trade and financial markets. Understanding these shifts requires awareness of global interdependencies and anticipation of cross-border economic impacts. International collaboration and dialogue can collectively address the challenges and opportunities of technological innovation, fostering a stable global economy. This approach encourages proactive management of volatility, paving the way for sustainable growth and prosperity.

Planetary Climate Oscillations

Embark on an exploration into the realm of planetary climate dynamics, where Earth's natural systems engage in a complex interplay of forces. Picture a world where ocean currents, atmospheric winds, and climate shifts work together like the components of a vast machine, each impacting the next in a finely tuned equilibrium. These fluctuations are not merely background elements in

Earth's climatic symphony; they are the maestros, directing weather patterns and shaping long-term climate trends. Central to this grand orchestration are interaction cycles that echo across time and space, influencing everything from daily weather to the future of ecosystems. The beauty of these systems lies in their intricacy, where seemingly minor changes can cascade through the environment, triggering a series of transformations affecting the entire planet.

As we delve deeper into this subject, the importance of comprehending these cycles becomes clear. Ocean currents act as the lifeblood of the Earth, redistributing heat and nutrients and molding climates across continents. Atmospheric variations, with their rhythmic sequences, serve as the Earth's breath, guiding the rise and fall of weather systems. Long-term climate changes unveil the planet's history and provide insights into its future, underlining the global repercussions of these cycles. By modeling these complex interactions, scientists can better understand both the predictability and unpredictability of climate behavior. Each subtopic exposes a layer of this intricate tapestry, inviting readers to uncover the dynamic interactions that govern our planet's climate and fostering a deeper appreciation for the seemingly small actions that drive global change.

The Role of Ocean Currents in Climate Regulation

The complex interplay of ocean currents is pivotal in Earth's climate regulation, functioning as a vast conveyor belt that redistributes heat globally. These currents arise from the combined effects of wind patterns, Earth's rotation, and variations in water density influenced by temperature and salinity. This dynamic system moderates climate by moving warm water from the equator toward the poles and cold water in the opposite direction. A prime example is the Gulf Stream, which significantly warms the North Atlantic, affecting European weather patterns. This exchange of oceanic heat impacts not only regional climates but also the global climate balance.

Recent research underscores the profound effects ocean currents have on climate variability, linking fluctuations to phenomena like El Niño and La Niña. These periodic changes in Pacific Ocean currents trigger global shifts in weather patterns, affecting precipitation, storm activity, and even agricultural productivity. Advances in satellite technology and modeling have shed light on these interactions, revealing how small changes in current strength or direction can lead to significant climatic consequences. Understanding these processes is essential for predicting future climate scenarios and developing strategies to mitigate negative impacts.

The interaction between ocean currents and atmospheric conditions forms a complex system where feedback loops can amplify or dampen climatic shifts.

A fascinating aspect is the role of ocean currents in the carbon cycle. Oceans absorb substantial atmospheric carbon dioxide, facilitated by water movement. Variations in current intensity can alter carbon uptake rates, affecting global carbon budgets and climate change trajectories. This is a vibrant area of research, with scientists exploring how changes in ocean dynamics might influence carbon sequestration amid rising global temperatures.

As climate evolution continues, understanding potential long-term shifts in ocean currents becomes crucial. Changes like the potential slowing of the Atlantic Meridional Overturning Circulation raise concerns about future climate stability. This critical component of the global ocean conveyor belt is vital for regulating Northern Hemisphere climate. Recent studies suggest that human-driven climate change might weaken these currents, possibly leading to drastic climatic consequences. Such scenarios necessitate a reevaluation of climate models and an urgent call for adaptive strategies in climate policy and management.

Exploring the connections between oceanic processes and broader climate patterns deepens our appreciation of Earth's interconnected systems. Examining ocean currents provides insight into natural climate regulation and uncovers paths for informed intervention. Encouraging reflection on these relationships can spark innovative thinking on harnessing or adapting to oceanic forces for climate resilience. Engaging with these ideas invites a broader exploration of how subtle changes in one part of the system can ripple through the entire planetary climate, promoting a proactive approach to understanding and addressing the challenges of a changing world.

Understanding Atmospheric Oscillations and Weather Patterns

Atmospheric fluctuations play a pivotal role in shaping our planet's climate, acting as dynamic cycles that drive weather variability across time. The El Niño-Southern Oscillation (ENSO) serves as a prime example, illustrating how changes in oceanic temperatures can shift atmospheric pressures, thus influencing global weather patterns. During an El Niño phase, warmer Pacific waters disrupt typical trade winds, resulting in increased rainfall in some areas while causing droughts in others. These fluctuations aren't isolated events; they have extensive impacts, highlighting the interconnected nature of Earth's climate systems. Recent research has shed light on these patterns, offering greater insights into their causes and effects, and improving our ability to predict them accurately.

Exploring further into atmospheric behaviors, scientists have identified a range of oscillatory patterns occurring at various timescales and intensities. The

North Atlantic Oscillation (NAO), for example, demonstrates a seesaw-like shift in atmospheric pressure between Iceland and the Azores, which affects weather conditions across Europe and North America. A positive NAO phase generally brings milder winters to Europe, while a negative phase results in colder, stormier seasons. Understanding these patterns provides valuable insights into regional climate trends and helps societies prepare for and adapt to potential changes, reducing risks to agriculture, infrastructure, and public health.

Advancements in climate science have integrated satellite data and sophisticated modeling techniques, granting unprecedented insights into these atmospheric patterns and their effects on weather. These tools unravel the complexities of atmospheric interactions, allowing for a more detailed understanding of feedback mechanisms that drive climate variability. By simulating possible scenarios, researchers can examine how both natural and human factors might influence these fluctuations, paving the way for more informed climate policies and strategic planning. This approach emphasizes the necessity of interdisciplinary collaboration and innovation in addressing the complex challenges posed by climate dynamics.

A key aspect of understanding atmospheric fluctuations is recognizing the balance between natural variability and human-induced changes. While natural phenomena like ENSO and NAO are inherent, human activities, such as greenhouse gas emissions, can intensify or weaken these patterns. This interplay prompts important questions about the future of climate fluctuations in a rapidly changing world. Scientists are now exploring how rising global temperatures could affect the frequency and intensity of these patterns, with consequences for long-term weather and climate stability. Engaging with these questions encourages readers to reflect on the broader implications of their actions on the global climate.

Applying insights from atmospheric fluctuations to practical applications offers significant benefits across multiple sectors. For instance, by understanding these patterns, agricultural planners can optimize planting schedules, and urban developers can design infrastructure resilient to extreme weather. Encouraging individuals and communities to leverage this knowledge fosters a proactive approach to climate adaptation. By aligning small-scale actions with broader climate patterns, readers are empowered to contribute meaningfully to addressing the challenges of climate variability, creating a ripple effect of positive change worldwide.

Long-Term Climate Shifts and Their Global Impacts

FEEDBACK LOOPS ACROSS SCALES 37

Long-term climate fluctuations are driven by intricate interactions among natural systems, evolving over extensive timescales. These changes stem from a mix of influences, such as variations in solar radiation, volcanic eruptions, and tectonic activity. A striking example is the alternation between ice ages and warmer interglacial periods. Recent research highlights how slight changes, like shifts in Earth's orbit and axial tilt, can subtly redistribute solar energy, leading to significant climate transformations. Paleoclimatology, through ice core and sediment analysis, uncovers patterns echoing through Earth's history, offering insights into current trends and future projections.

Today, the rapid rate of human-induced climate change adds complexity to these natural cycles. Activities like burning fossil fuels and deforestation have elevated greenhouse gas concentrations to unprecedented levels. This human influence intensifies natural processes such as the albedo effect and carbon cycle dynamics, hastening warming trends and disturbing established climate patterns. For example, the melting of polar ice caps contributes to rising sea levels and decreases Earth's reflectivity, further accelerating global warming—a chain reaction with worldwide impacts.

Oceanic and atmospheric circulation patterns play a vital role amid these changes. The Atlantic Meridional Overturning Circulation (AMOC) is crucial in regulating climate by moving heat between the equator and polar regions. Recent findings suggest the AMOC might be weakening, potentially causing regional climate disruptions, altered rainfall patterns, and more extreme weather. Understanding these dynamics underscores the importance of sophisticated climate models that incorporate interaction cycles and nonlinear processes, providing accurate forecasts for decision-makers.

Emerging technologies are enhancing our ability to model and predict these complex interactions. Machine learning, for instance, analyzes vast datasets, revealing hidden patterns and offering more precise predictions. By simulating various scenarios, these technologies provide actionable insights for developing adaptive strategies to mitigate adverse effects and capitalize on new opportunities. Integrating indigenous knowledge, which offers unique perspectives on climate variability, further enriches our understanding and response strategies.

To empower individuals and communities, it is crucial to translate scientific findings into practical actions. Building resilience involves promoting sustainable practices like regenerative agriculture and embracing renewable energy, which not only cut emissions but also bolster local ecosystems. Encouraging cross-disciplinary collaboration can spark innovative solutions, driving systemic change. Reflecting on the interconnectedness of these systems fosters a deeper appreciation of how every action impacts the global climate narrative, urging intentional and forward-thinking action.

Modeling Complex Climate Interactions Through Feedback Loops

Grasping the complexities of planetary climate variations involves exploring the intricate network of interaction cycles that govern these natural systems. Earth's climate operates much like a complex orchestra, where every component has a significant role, and the interaction cycles act as conductors, directing the symphony. These cycles intricately weave through the planet's climate fabric, influencing the connections between the atmosphere, oceans, and land ecosystems. By studying these dynamics, we can uncover their substantial influence on climate stability and changes, providing insights into the forces that drive long-term climate shifts.

Recent breakthroughs in climate science have underscored the importance of interaction cycles in simulating complex relationships. For example, Earth's surface reflectivity, or albedo, serves as a crucial interaction cycle. As polar ice diminishes due to rising temperatures, decreased reflectivity increases solar absorption, further warming the planet and speeding up ice melt. This positive cycle demonstrates how interconnected systems can magnify changes, leading to significant climate transformations. Researchers use advanced models to simulate these processes, offering a glimpse into potential future climate scenarios.

The introduction of machine learning and artificial intelligence in climate modeling represents a transformative leap in understanding interaction cycles. These technologies allow scientists to process large datasets, discover hidden patterns, and predict outcomes with exceptional precision. By integrating machine learning algorithms with traditional climate models, researchers can refine predictions and identify emerging phenomena that might otherwise remain unnoticed. This combination of cutting-edge technology with climate science promises more accurate forecasts, enabling policymakers to make informed decisions to mitigate climate change effects.

A notable illustration of interaction cycles in action is the El Niño-Southern Oscillation (ENSO), a periodic change in sea surface temperatures across the equatorial Pacific Ocean. ENSO showcases how atmospheric and oceanic interaction cycles can drive global climate patterns, affecting weather extremes such as droughts and floods. By understanding ENSO's mechanisms, scientists can develop predictive models to anticipate its occurrence and impact, helping societies better prepare for associated challenges. This knowledge highlights the importance of studying interaction cycles to comprehend broader climate variability implications.

As we explore the complex web of planetary climate variations, it becomes evident that interaction cycles are not just abstract ideas but crucial forces shaping our world. The insights gained from modeling these interactions are key to developing strategies for sustainable climate management. By fostering a deeper understanding of interaction cycles and their role in climate systems, individuals and communities can engage in informed actions that contribute to global resilience. Encouraging a holistic perspective on climate interactions, these insights inspire a sense of stewardship, empowering readers to actively participate in the quest for a balanced and thriving planet.

Navigating the complex web of response loops reveals their significant role in shaping our environment. These self-sustaining cycles are evident everywhere—from the minute interactions in molecular reactions to the large-scale fluctuations of economies and climate systems. Such mechanisms highlight how seemingly small changes can escalate into monumental shifts, underscoring the intricate connectivity of various systems. By decoding these patterns, we gain valuable insights into how minor interventions might ripple across broader contexts, influencing larger outcomes. This chapter brings to light the elegance of these interaction cycles, encouraging us to acknowledge their impact in everyday systems. As we journey further into the universal language of patterns, we are prompted to consider how intentional actions can harness these dynamics for positive transformation. What initially appears as an isolated action can, through the power of these cycles, spark significant change, urging us to reflect on how our contributions might resonate within the wider fabric of the world.

Chapter Three
Network Effects And Emergence

As we delve into the intricate web of connections that shape our universe, we embark on a journey to uncover the marvels of network dynamics and spontaneous organization. Picture yourself at the boundary of a sprawling forest, where each tree symbolizes a point in a vast web, its branches extending to meet countless others. The dance of sunlight and shadows among the leaves reflects the intricate relationships that govern our world. These connections, whether found in the neural circuits of our minds or the expansive arms of galaxies, reveal a profound insight: the collective is far more powerful than the individual elements. By examining these networks, we gain insight into the subtle forces that drive natural order and the remarkable ability of systems to self-organize.

In the depths of our minds, neural pathways chart the complex corridors of thought and perception. Every synapse and neuron forms a crucial link in the delicate architecture of consciousness. These microscopic exchanges lay the groundwork for our capacity to learn, adapt, and innovate. Meanwhile, in the tapestry of human society, connections and influences ripple through communities, subtly shaping our collective behaviors. These social webs, much like the underground networks that sustain a forest, are essential to the growth of ideas and movements, often branching out in unexpected ways with profound consequences.

Beyond our immediate view, the vast expanse of galaxies reminds us of the grand scale of network effects. The gravitational pull of stars, gas, and dark matter creates magnificent structures, showcasing the power of interconnectedness throughout the cosmos. As we explore these complex systems, we unlock insights into the fundamental processes that govern both our world and the universe. This chapter invites you to discover how understanding network dynamics and emergence can illuminate the forces that shape everything from the cognitive to the cosmic.

Neural Networks and Brain Function

Understanding the influence of network effects on brain function begins with marveling at the intricate complexity of neural systems. These elaborate networks, resembling a sprawling tapestry of linked pathways, form the bedrock of our thoughts, feelings, and behaviors. The brain, an exemplar of nature's brilliance, orchestrates a dynamic symphony of neurons communicating through billions of synapses. This exchange of signals and responses is fluid and transformative, constantly adapting to new experiences and information. At the core of this adaptability is neural plasticity, the brain's extraordinary ability to reorganize itself by forging new connections. As neurons engage in continuous interaction, they generate a cascade of electrical impulses that sculpt our cognitive landscape, affecting everything from memory to perception.

Delving into the realm of neural networks unveils a world where simple interactions lead to remarkable capabilities. Within this complex web, feedback loops are crucial in refining and enhancing cognitive functions. These loops, functioning like nature's self-correcting systems, enable the brain to learn, adapt, and optimize its performance. As we explore the emergent properties of complex brain functions, we discover how these interactions culminate in the sophisticated abilities that characterize human intelligence. The interplay of these processes highlights the brain's capacity to synthesize vast information into coherent patterns, allowing us to navigate and interpret our surroundings. This exploration not only reveals the wonder of brain function but also prepares us to understand how similar effects manifest in broader contexts, guiding us into discussions of social and cosmic systems.

Understanding Neural Plasticity in Network Formation

Neural plasticity, the brain's extraordinary capability to adapt and reorganize, is a pivotal focus in modern neuroscience. This adaptability involves the continual formation and modification of synaptic connections, essential for learning, memory, and recovery from brain injuries. Recent research has revealed how these changes occur on both cellular and systemic levels. By creating new connections and eliminating less-used pathways, the brain optimizes its neural framework, demonstrating a biological efficiency that captivates scientists and innovators. This flexibility highlights the brain's resilience, providing insights into enhancing cognitive functions through targeted interventions.

Advancements in imaging technologies and computational models have unveiled the complex interactions of synapses during network formation. Techniques like functional MRI and diffusion tensor imaging offer unprecedented glimpses into the brain's evolving structure, capturing the shifts in neural pathways responding to new stimuli and experiences. These images depict not only the physical restructuring of neural circuits but also the mechanisms driving these changes. The coordination of neurotransmitters, growth factors, and electrical activity orchestrates this intricate adaptation process, inspiring scientific exploration and practical applications in neurorehabilitation and artificial intelligence.

The concept of neural plasticity influences cognitive development theories and educational strategies. Understanding how experiences shape neural architecture prompts a reevaluation of learning environments, emphasizing rich, varied stimuli to foster cognitive growth. This perspective supports adaptable, personalized educational systems aligned with each learner's unique neural profile. Furthermore, principles of plasticity are instrumental in developing therapies for neurodegenerative diseases, offering hope for interventions that may slow or reverse cognitive decline by leveraging the brain's inherent adaptability.

In artificial intelligence, insights from neural plasticity fuel innovations in machine learning and neural network design. By emulating the brain's adaptive processes, engineers develop algorithms that learn and evolve similarly to human cognition. This exchange of ideas bridges biological and artificial networks, leading to more intuitive, efficient AI systems. Applying these principles in technology not only enhances machine capabilities but also enriches our understanding of human cognition, creating a feedback loop of knowledge transfer benefiting both fields.

The brain's adaptability invites intriguing questions about the future of cognitive enhancement and human potential. Could we someday deliberately tap into plasticity to expand our intellectual horizons or recover lost abilities? This possibility transforms our understanding of human limits, suggesting a future where biological and artificial neural networks collaborate to unlock new dimensions of potential. As we explore these frontiers, the study of neural plasticity promises to illuminate the mind's mysteries and empower individuals to shape their cognitive destinies, testifying to the profound interconnectedness of all systems.

Mapping the Synaptic Connections and Information Flow

Neural pathways in the brain form the essential routes for information transmission, crafting the complex framework of cognition. Central to this dynamic activity is synaptic connectivity, where neurons establish links to facilitate signal transmission. Each synapse functions as a gateway, enabling the flow of electrical impulses fundamental to thought, memory, and awareness. This intricate web of interactions is perpetually evolving, adapting and reorganizing in response to internal stimuli and external experiences. The brilliance of this system lies in its ability to balance efficiency with adaptability, allowing the brain to seamlessly incorporate new information while preserving existing knowledge structures.

Recent breakthroughs in neuroimaging and computational modeling have detailed the astonishing architecture of these synaptic connections. Techniques like diffusion tensor imaging (DTI) and functional magnetic resonance imaging (fMRI) allow researchers to visualize the brain's connectivity in vivid detail, outlining the pathways that underpin mental processes. These tools emphasize the importance of local circuits for specific tasks and global networks for integrating information across various brain regions. This dual approach ensures rapid processing and response to stimuli, vital for survival and adaptation in a complex world.

Synaptic connections are not merely conduits of information; they embody neural plasticity, the brain's ability to modify connections based on experience and learning. This adaptability is key to acquiring new skills and recovering from injuries. Synaptic plasticity, the process by which synapses strengthen or weaken over time, is influenced by activation frequency and neurotransmitter release. Understanding this process has profound implications for treating neurological disorders and enhancing cognitive function. Emerging therapies, such as transcranial magnetic stimulation (TMS) and neurofeedback, aim to harness this plasticity, offering potential avenues for treating conditions like depression and schizophrenia.

A captivating aspect of synaptic information flow is the role of feedback mechanisms in refining cognitive processes. Feedback loops, both positive and negative, regulate synaptic strength and network stability. Positive feedback can amplify signal strength, reinforcing learning and memory, while negative feedback maintains balance, preventing overexcitation that could lead to disorders like epilepsy. This complex interplay of feedback mechanisms highlights the brain's ability to self-regulate and optimize performance, showcasing a remarkable level of sophistication and self-awareness.

Applying these insights to artificial intelligence, understanding synaptic connectivity can inform the design of neural networks and machine learning algorithms. By emulating the brain's structure—emphasizing adaptability, feedback, and network coherence—engineers can develop smarter, more efficient AI systems. This convergence of neuroscience and technology not only advances our understanding of the brain but also inspires innovative solutions to complex challenges. Encouraging interdisciplinary collaboration could lead to breakthroughs that enhance both human and artificial intelligence, deepening our comprehension of the universe's most intricate framework.

The Role of Feedback Loops in Cognitive Processes

Feedback loops are vital for understanding how the brain evolves and adapts. These mechanisms enable real-time information processing, allowing the brain to learn from experiences and modify its responses. A classic example is the refinement of motor skills through sensory input processing. When learning a musical instrument, for example, feedback loops enable the brain to correct errors by comparing intended actions with actual results, enhancing skill over time.

The complexity of feedback systems is highlighted by the brain's reward circuitry, which employs neurotransmitters like dopamine to reinforce advantageous behaviors. When an action yields a positive result, dopamine strengthens relevant synaptic connections, increasing the likelihood of repeating the behavior. This biological reinforcement mirrors how artificial neural networks use feedback loops to refine training algorithms for better performance.

Recent research has shed light on how feedback loops contribute to complex cognitive abilities like decision-making and problem-solving. Functional magnetic resonance imaging (fMRI) studies reveal how various brain regions communicate through feedback loops to evaluate options and predict outcomes. Understanding these processes can lead to better interventions for cognitive impairments, aiding individuals with conditions like ADHD or autism, where feedback processing may differ.

Feedback loops also play a role in collective intelligence, as the exchange of ideas within social networks fosters emergent properties and shared knowledge. In education, feedback loops between teachers and students create an environment of continuous learning. Teachers adapt their strategies based on student feedback, while students deepen their understanding through iterative input, resulting in improved educational outcomes.

To leverage feedback loops in personal and professional settings, one can adopt iterative learning and adaptation strategies. This might include regular self-assessment, seeking feedback from peers, or using tools to monitor progress. By understanding feedback mechanisms, individuals can create environments that encourage growth, creativity, and innovation, contributing to broader systemic change. Consider implementing a feedback loop to achieve a specific goal and identify necessary adjustments to optimize the process.

Emergent Properties of Complex Brain Functions

NETWORK EFFECTS AND EMERGENCE 45

Complex brain functions reveal remarkable emerging characteristics, where intricate systems give rise to behaviors and attributes not easily predicted by studying individual components. Central to this complexity is the brain's capacity to integrate vast information through its intricate web of neurons. These neural networks process sensory inputs, generate thoughts, and drive actions through dynamic interactions and adaptive learning. Neural plasticity, the brain's ability to reorganize through new synaptic connections, is foundational to these emerging properties, allowing for constant refinement of processes, enhancing learning and memory with impressive efficiency.

Advancements in neuroimaging and computational modeling have deepened our understanding of the brain's dynamic functions. These technologies enable researchers to observe brain activity in real-time, illustrating how different regions collaborate to generate complex behaviors. Notably, studies have identified brain frameworks responsible for higher cognitive functions, such as the default mode network, active during introspective thought and imagination. Understanding these interactions helps unravel the neural basis of consciousness and creativity, offering significant implications for fields like artificial intelligence and education.

Feedback loops are crucial in shaping cognitive processes, continuously refining neural activity to ensure appropriate responses to stimuli. In decision-making, for example, the brain evaluates past experiences and anticipates outcomes, adjusting strategies as needed. This iterative process acts as a self-correcting mechanism, enhancing cognitive accuracy and efficiency. Studying these feedback systems allows researchers to develop targeted interventions for neurological disorders, potentially transforming therapeutic approaches.

Exploring the brain's emerging properties also involves considering the impact of environmental factors and experiences. The interaction between genetic predispositions and external stimuli contributes to the diversity of cognitive abilities across individuals. This idea challenges the traditional view of fixed intelligence, suggesting instead that cognitive potential is adaptable and can be nurtured through enriched environments and targeted training. The educational implications are significant, highlighting the potential for personalized learning approaches tailored to individual strengths and weaknesses.

To apply these insights effectively, fostering environments that stimulate cognitive growth and adaptability is key. Encouraging activities that enhance problem-solving, critical thinking, and creativity can expand the brain's capacity for new properties. Additionally, mindfulness practices and cognitive exercises support neural plasticity, promoting mental well-being. By applying these strategies, individuals and communities can tap into the transformative power of brain functions, driving personal and societal progress.

Social Connection Patterns

Picture a vast web of connections that intertwines us all—a complex tapestry composed of myriad interactions and influences. At its heart, this network reveals a wealth of potential far greater than the sum of its parts. Each point in this web—every greeting, discussion, and shared moment—acts as a channel through which ideas and behaviors ripple outward, echoing like waves across a pond. This chapter explores the intriguing dynamics of these interactions, illustrating how even the smallest exchanges can spark significant changes in societal structures. From the close bonds within tight-knit communities to the expansive, cross-continental links, the true power of social frameworks lies in their capacity to amplify and hasten transformation.

As we traverse this intricate landscape, the phenomenon of social contagion becomes apparent, demonstrating how ideas and behaviors spread autonomously through networks. The stability of these systems often depends on the strength of weak ties—those seemingly minor connections that create bridges between diverse groups. These ties bolster the resilience and adaptability of the entire network, enabling it to endure and evolve in the face of challenges. Within these intricate webs, new patterns and innovations emerge, often unexpectedly, as collective interactions give rise to novel behaviors. Society's innovative capacity derives from the collective intelligence harnessed within these systems, providing a rich source of creativity and solutions. Through examining these facets, the transformative potential of social connections becomes clear, inviting a deeper understanding of how our interconnectedness can drive positive change.

Mapping the Dynamics of Social Contagion

Social contagion describes how behaviors, ideas, and emotions spread across societies much like viruses. This process parallels how memes and trends proliferate, highlighting the influence of social networks on collective thought. Unlike traditional communication, social contagion operates through the complex web of personal connections, where influence is often indirect or subtle. Studies reveal the intricate nature of these interactions, showing how isolated actions can spark widespread behavioral change. This phenomenon underscores the power of small nudges that ripple through networks, turning individual decisions into significant societal trends. It prompts us to reconsider our community engagement, recognizing that each interaction can contribute to broader change.

Weak ties are crucial to the strength and flexibility of social networks. While strong ties—like those with close friends or family—offer emotional support, weak ties—connections with acquaintances or distant colleagues—act as conduits for new ideas and opportunities. These connections are crucial for spreading innovative concepts and fostering creativity, often linking disparate groups and bringing in fresh perspectives. Research on weak ties reveals their surprising role in maintaining network resilience and adaptability. Building diverse connections can enhance access to valuable resources and create environments conducive to creative breakthroughs and collaborative problem-solving.

Emergent behavior in complex social networks demonstrates how intricate patterns arise from simple actions. This is similar to the murmuration of starlings or the synchronized flashing of fireflies, where local interactions lead to coherent group dynamics. In human societies, emergent behavior appears in various forms, from spontaneous crowd movements to the rapid adoption of new technologies. Understanding these properties enables us to harness collective potential for constructive goals. By fostering open communication and adaptive learning, we can unlock latent community capabilities, paving the way for more effective collaboration.

Collective intelligence epitomizes networked collaboration, where the combined expertise of individuals creates solutions beyond the reach of isolated efforts. This synergy is evident in crowdsourced platforms and open-source projects, where diverse contributions lead to innovation. The challenge is directing this collective potential toward beneficial objectives. By utilizing digital tools and encouraging inclusive participation, we can amplify the impact of collective intelligence, transforming how societies tackle complex challenges. This perspective encourages viewing networks not just as information channels, but as dynamic systems capable of producing transformative insights and solutions.

To effectively leverage social contagion, individuals and organizations can employ strategies that enhance positive influences within networks. By consciously spreading constructive behaviors and ideas, one can trigger beneficial cascades. Identifying key influencers—those with significant reach or impact—can expedite this process, acting as catalysts within the network. Encouraging diverse connections, fostering trust and open dialogue, and using digital platforms to expand reach are practical steps to amplify positive social contagion. These efforts highlight the potential for mindful engagement to create ripples of change, breaking conventional barriers and fostering a more interconnected and resilient society.

The Role of Weak Ties in Network Resilience

In the complex network of social connections, weak ties are the subtle yet essential strands that add resilience to human relationships. These seemingly minor links, often marked by occasional interactions and low emotional depth, are vital for the strength of social systems. Unlike the traditional emphasis on strong ties—those deep, close relationships—weak ties act as channels for new information and varied perspectives. By connecting different social groups, they facilitate the spread of ideas and resources, enhancing a community's adaptability and resilience. This perspective on weak ties challenges conventional thinking, offering a new view on social connectivity dynamics.

Recent studies in social network theory reveal the unexpected power of weak ties in promoting societal resilience. Sociologist Mark Granovetter's research highlights their significance in job searches and knowledge sharing, showing that opportunities often come not from close friends but from acquaintances. This concept, known as "the strength of weak ties," demonstrates how indirect connections can drive significant personal and professional growth. By linking otherwise isolated groups, weak ties enable networks to endure disruptions, strengthening their overall structure against external challenges.

In the realm of digital platforms and online interactions, weak ties become even more significant. Social media illustrates how these connections can multiply, facilitating rapid information exchange and fostering global movements. Platforms like Twitter and LinkedIn use weak ties to amplify voices and connect people across vast distances and cultural divides. As users engage with a wider range of contacts, they gain access to a broader array of ideas, boosting their cognitive diversity and problem-solving capabilities. This highlights how digital weak ties can create a resilient and innovative social structure, capable of adapting to a constantly changing environment.

Strategically nurturing weak ties offers practical insights for enhancing community and organizational resilience. Creating environments that value and facilitate diverse interactions can strengthen social structures. For example, organizations can host collaborative spaces or events that encourage cross-departmental interaction, fostering weak ties that can lead to innovative solutions. Similarly, community initiatives that unite diverse groups through shared goals can create robust networks that thrive on the strength of weak ties.

Considering the role of weak ties prompts reflection on their broader implications for societal innovation and transformation. Can the power of weak ties be harnessed to tackle global challenges like climate change or economic inequality? By fostering networks that value diversity and openness, individuals and organizations might unlock unprecedented potential for creative solutions and collective intelligence. As we navigate an increasingly interconnected world, the insights gained from understanding weak ties could guide the development of resilient, adaptive societies capable of addressing modern complexities.

Emergent Behavior in Complex Social Networks

Emergent behavior in intricate social systems is a fascinating aspect of interconnected frameworks where individual actions and interactions give birth to new, often surprising, collective phenomena. This behavior is evident in the swift dissemination of ideas, trends, or innovations within societies, spreading rapidly across both digital and physical realms. Social frameworks, whether offline or online, display a complex web of interactions that cannot be wholly predicted by simply summing up their parts. This phenomenon challenges traditional linear perspectives, highlighting the need to appreciate the nuanced, often nonlinear dynamics that define complex systems. Understanding these emergent properties requires a detailed examination of the subtle rules and interactions that guide individual behavior and collective outcomes.

Recent progress in network theory has highlighted the pivotal role of certain nodes, or individuals, in facilitating these emergent behaviors. Key figures within a framework, often referred to as influencers or central nodes, act as catalysts for widespread change by amplifying messages and ideas through their extensive connections. This amplification effect emphasizes the importance of network structure—the arrangement of connections—where minor changes can lead to significant impacts. Studying these influential nodes provides valuable insights into how social movements gain momentum, misinformation spreads, and collective action is coordinated. Researchers continue to explore these dynamics, using advanced models and simulations to predict and harness emergent behavior for societal benefit.

In analyzing intricate social structures, the interplay of weak and strong ties must be considered. Weak ties, though seemingly minor, serve as critical bridges between disparate groups, enabling the flow of information and innovation across network boundaries. This bridging function can spark the emergence of novel ideas and collaborations, as it provides access to diverse perspectives and resources. By strategically nurturing these weak ties, individuals and organizations can create an environment conducive to innovation and adaptability. Balancing close-knit communities and broad connections is crucial for cultivating a resilient and dynamic system capable of emergent behavior.

The study of emergent behavior in such systems also highlights the strength of collective intelligence, where the aggregation of individual knowledge and insights surpasses the capabilities of single entities. This phenomenon is evident in platforms that leverage crowdsourcing and collaborative problem-solving, tapping into the wisdom of diverse groups to tackle complex challenges. Harnessing this collective intelligence requires deliberately designing systems that promote open communication, trust, and inclusivity, ensuring that all voices are heard and valued. By understanding the principles of emergent behavior, we can establish environments that not only anticipate but also thrive amidst complexity and change.

To translate these insights into practical applications, individuals and organizations can adopt strategies that enhance connectivity and resilience within systems. Encouraging diversity within networks, fostering transparent communication, and cultivating environments that embrace adaptability are key steps toward harnessing emergent behavior for beneficial outcomes. By recognizing the potential of interconnected frameworks and the emergent phenomena they produce, we can better navigate the complexities of modern society, driving innovation and fostering a culture of continuous growth and learning. This perspective empowers individuals and equips communities to address global challenges with agility and creativity, making the invisible forces of networks visible and actionable.

Harnessing Collective Intelligence for Societal Innovation

In the sphere of societal innovation, collective intelligence represents a compelling framework for unleashing the cognitive power of groups. This shared intelligence emerges from the synergy of individuals working together, often achieving results beyond the capabilities of any single participant. In today's digitally connected environment, the potential for collective intelligence to foster societal transformation is vast, especially as digital tools open new paths for cooperation. To unlock this potential, we must understand how groups communicate, solve problems, and make decisions. Modern tools like digital platforms and data-driven insights can streamline these processes, empowering diverse groups to tackle complex issues in unprecedented ways.

A striking example of collective intelligence is the open-source software movement. Initiatives such as Linux and Apache illustrate how a dispersed community of developers can collaboratively produce robust, innovative software solutions, outperforming traditional approaches in both scale and speed. These projects thrive on global contributions, with each member bringing distinct skills and viewpoints. By encouraging an open exchange of ideas, these initiatives harness the collective wisdom of their communities to evolve and improve continuously. This approach can extend beyond software, providing a template for addressing societal challenges like climate change, public health, and education reform, where diverse perspectives and rapid iteration are crucial.

The advent of crowdsourcing platforms further highlights the potential of collective intelligence. Sites like Kickstarter and Indiegogo have transformed project funding by leveraging the power of the crowd to identify and support novel ideas. These platforms democratize funding access, allowing people from varied backgrounds to pool resources for projects they believe in. This not only empowers entrepreneurs but also generates a feedback loop, enabling creators to assess public interest and refine their concepts. Insights from these interactions can drive larger-scale innovations, reflecting the broader community's needs and values.

As we explore collective intelligence, the role of diversity in enhancing group performance becomes clear. Research consistently shows that diversity in thought, experience, and background leads to more innovative solutions and better decision-making. By intentionally cultivating diverse teams and networks, organizations can tap into a broader range of perspectives, resulting in more robust and creative outcomes. This approach not only enriches the innovation process but also ensures that the solutions developed are more inclusive and equitable, addressing the needs of a wider spectrum of society.

To effectively harness collective intelligence for societal innovation, creating environments that foster open communication and constructive dissent is essential. Encouraging a culture where individuals feel empowered to share ideas and challenge assumptions can lead to breakthroughs that might otherwise be overlooked. Leveraging technology to facilitate these interactions can bridge geographical and cultural divides, amplifying the reach and impact of collaborative efforts. By nurturing spaces where collective intelligence can thrive, society can unlock new possibilities for innovation, driving transformative change across various domains. Readers are encouraged to consider how they can contribute to and benefit from these collaborative efforts, recognizing the profound impact collective intelligence can have on shaping a better future.

Galactic Structure Formation

In the vast expanse of the universe, galaxies engage in a captivating interplay, forming a complex mosaic that reveals the intricate design of the cosmos. Their creation and transformation highlight the sophisticated patterns that define our universe, echoing the network effects seen in both human society and the infinite stretches of space. As stars, gas, and dark matter converge, they craft the awe-inspiring structures we know as galaxies. These formations are not mere random collections of celestial objects but are sculpted by fundamental laws of connectivity and interaction. Central to this celestial symphony is dark matter, a mysterious yet pervasive force that orchestrates the gravitational dynamics crucial to galaxy formation. Its invisible influence shapes the mass distribution that determines how galaxies cluster and evolve, providing a framework for the visible matter to gather and form the dazzling shapes that intrigue us.

Whether spiral or elliptical, galaxies display unique configurations that reveal the underlying forces at work. By examining these stellar arrangements, we uncover the principles of emergence that guide their development, similar to the patterns found in neural networks and social systems. Galaxies are intricately linked through gravitational forces, drawing them into clusters and larger systems, reminiscent of the connections that bind communities and ideas. This vast network, often called the cosmic web, stretches across the universe, illustrating the profound interconnection of all galactic formations. In this exploration, we find not only the wonders of space but also reflections of universal patterns that influence all forms of existence, from the tiniest particles to the largest astronomical structures. As we delve into these celestial phenomena, the principles of emergence and network effects become apparent, revealing the universe's inherent drive towards organized complexity.

The Role of Dark Matter in Galactic Formation

Dark matter, shrouded in mystery, is essential to the grand design of universe formation. Although it cannot be seen through telescopes due to its non-interaction with electromagnetic forces, its gravitational pull is critical. It forms a framework that holds galaxies together, acting as an adhesive force. Recent models illustrate how dark matter forms halos, which in turn draw in the baryonic matter that composes stars and planets. These halos lay the groundwork for galaxies, influencing both their birth and evolution. The relationship between dark and visible matter highlights the unseen elements that mold our universe, urging us to reconsider cosmic structures.

Observations from cosmic microwave background radiation and galaxy rotation curves emphasize dark matter's importance in explaining the universe's architecture. Without it, the speed at which galaxies rotate suggests they lack the mass necessary to remain intact. This discrepancy between observed dynamics and visible mass distribution supports the theory that dark matter constitutes nearly 27% of the universe's mass-energy content. Such insights have driven astrophysicists to refine cosmological models, recognizing dark matter as a cornerstone of cosmic theory. The interaction between dark and baryonic matter shapes not only individual galaxies but also large-scale formations like galaxy clusters, creating a complex network of cosmic interconnections.

Current research is delving into the nature of dark matter, investigating its possible forms and properties. Theories range from weakly interacting massive particles (WIMPs) to axions, each offering different implications for galaxy formation. Advanced experiments, both underground and in space, aim to detect the elusive dark matter particles, potentially revealing more about their nature. These scientific pursuits exemplify the quest to decode the universe's hidden mass, fostering technological innovation and cross-disciplinary collaboration. As our understanding of dark matter deepens, so does our appreciation of its role in shaping the cosmos.

Despite the progress, the true nature of dark matter remains one of astrophysics' deepest questions. This mystery invites diverse theories, from physicists proposing new particles to astronomers studying gravitational lensing effects hinting at dark matter. Some researchers have even explored alternative explanations, such as changes to gravitational laws on a cosmic scale. These debates enrich our discussions, encouraging a dynamic exchange of ideas that stretches the limits of current knowledge. By embracing various perspectives, we widen our understanding and create a fertile ground for discovery.

Reflecting on the cosmic importance of dark matter, one is reminded of the patterns that govern the universe. Dark matter's invisible influence orchestrates galaxy formation, showing how unseen forces can profoundly affect what is observable. This invites contemplation of the universe's interconnectedness, from the tiniest particles to the vastest galactic structures. As we explore dark matter's mysteries, we are encouraged to consider the broader implications for astronomy and our understanding of the universe and our place within it. This cosmic journey offers insight into the universe's complexity and beauty, inspiring wonder and curiosity in those seeking its secrets.

Spiral and Elliptical Galaxies: Patterns of Emergence

In the intricate cosmos, spiral and elliptical galaxies emerge as stunning cosmic formations shaped by gravitational dynamics and cosmic material. Spiral galaxies, with their elegant, curving arms, are known for active star formation in their youthful regions. These structures likely result from density waves traversing the galactic disk, compressing interstellar gas and sparking star creation. This process forms a captivating cycle where new stars invigorate the galaxy, promoting further wave activity. This cosmic choreography exemplifies how localized interactions can lead to grand designs, demonstrating the power of minor actions to drive significant transformations.

Elliptical galaxies, on the other hand, tell a unique story through their formation and evolution. These vast, spheroidal entities primarily consist of older stars and show minimal new star formation. They are believed to form through mergers of smaller galaxies, a process that disrupts stellar orbits, leading to a more randomized, spherical star arrangement. Despite the chaos of these mergers, they eventually result in stable structures, illustrating how order can emerge from chaos, a concept reflected in both natural and social systems.

Recent advances in astronomy have shed light on these galactic structures. Using tools like the Hubble Space Telescope and ALMA, researchers have uncovered new insights into these galaxies' composition and behavior. For example, supermassive black holes at the centers of both galaxy types influence their evolution, affecting star formation rates and structure. These findings highlight the complex feedback mechanisms shaping galaxies, mirroring the universe's intricate complexity.

The patterns in spiral and elliptical galaxies prompt reflection on broader principles of emergence and self-organization. They show how simple principles can create complex systems, resonating with concepts in other fields like neural networks or social systems, where individual elements collectively generate sophisticated behaviors. Studying these celestial formations offers a deeper understanding of cosmic phenomena and appreciation for universal principles that apply across scales, from the micro to the monumental.

Considering the broader implications of these galactic patterns, one might wonder how these cosmic insights can inform strategies for positive change within human systems. The emergent properties of galaxies, resulting from myriad small interactions, suggest that mindful micro-level adjustments can lead to significant systemic changes. By learning from the cosmos, individuals and communities can be inspired to engage thoughtfully with their environments, recognizing the potential for small, intentional actions to spark larger, beneficial transformations.

Gravitational Interactions and Galactic Clustering

The cosmic ballet of galaxies, orchestrated by gravitational forces, is a captivating spectacle that shapes the universe. Central to this grand choreography is dark matter, an elusive entity that outweighs visible matter and forms the invisible framework around which galaxies cluster. Recent research indicates that these dark matter halos act as cosmic glue, drawing stars and gas to craft the complex patterns of galactic clusters. By studying these interactions, we gain insight into cosmic evolution and the delicate balance between visible and hidden forces.

Gravitational interactions are not only foundational but also transformative, amplifying tiny fluctuations in the early universe to create the vast structures we see today. These forces actively shape galactic destinies, pulling galaxies into clusters with varying densities. As galaxies are drawn together, they engage in a dynamic interplay, sometimes merging into larger systems or colliding spectacularly. This ongoing process demonstrates that galactic clustering is an evolving story, not a static occurrence.

Exciting advancements in astrophysics explore how gravity influences galaxy shapes and behaviors within clusters. Observations show that galaxies in dense environments often appear more elliptical, while those in less crowded areas maintain spiral forms. This suggests that the gravitational dynamics within clusters can alter a galaxy's structure and star-forming activities. Studying these interactions enriches our understanding of galactic life cycles and cosmic diversity.

Looking at the universe's vast web of interconnected galaxies, we see gravity's influence extending beyond individual clusters. This web-like structure illustrates matter's distribution on a grand scale and hints at the universe's expansion and dark energy's role. Viewing the universe as a unified system challenges us to rethink it as more than isolated entities.

For those seeking practical insights, exploring gravitational interactions and galactic clustering offers applications in various fields. Recognizing the universal nature of gravitational forces can inspire parallels with complex systems like social networks or ecosystems, where interactions drive collective behavior. Encouraging interdisciplinary research can lead to innovative approaches in understanding complex systems, akin to the cosmic dance of galaxies. This exploration invites us to consider how cosmic principles might illuminate solutions to earthly challenges.

Cosmic Web: Interconnectedness of Galactic Structures

The vast and intricate cosmic web presents a remarkable perspective on the universe's interconnected nature. At its essence, this web comprises dense clusters of galaxies connected by dark matter filaments, creating an immense structure spanning the cosmos. Recent advancements with telescopes and simulations have unveiled that these filaments serve as gravitational pathways, steering the formation and movement of galaxies. This discovery underscores dark matter's crucial role in shaping the universe's large-scale structure, acting as the unseen framework around which galaxy clusters form. As scientists explore this cosmic network further, they find patterns resembling neural networks, hinting at a universal model of connectivity that spans different scales.

When examining galaxy cluster formation within this web, these clusters emerge not as isolated entities but as parts of a dynamic ecosystem influenced by surrounding mass distribution. The gravitational pull among galaxies within a cluster leads to complex behaviors, where interactions result in matter coming together. This mirrors feedback loops found in biological and social systems, where individual interactions lead to larger identity-defining behaviors. Understanding these interactions sheds light on galaxies' evolutionary paths and provides a guide to deciphering matter and energy's intricate dance in the universe.

The cosmic web also fosters the development of distinct galaxy shapes, like spiral and elliptical galaxies. These shapes originate from initial conditions set by the web's structure and gravitational interactions. Spiral galaxies, with their twisting arms, often form in areas where tidal forces and rotational dynamics are strong, organizing stars and gas. In contrast, elliptical galaxies usually arise in denser regions where frequent mergers create a more uniform appearance. This variety in galaxy forms highlights the cosmos's adaptability, where different initial conditions and interactions create a rich array of structures across the universe.

As cosmologists map the cosmic web, they discover surprising parallels with complex systems on Earth. The distribution of galaxies along the web's filaments resembles how social networks or power grids organize around central hubs, each node affecting the entire system's flow and connectivity. This suggests that principles of connectivity and emergence are universal, offering lessons in resilience and adaptability applicable to human-designed systems. By studying these cosmic connections, we gain insights into creating robust and sustainable networks in technology, society, and the environment.

Exploring the cosmic web invites contemplation of interconnectedness's implications for future cosmic exploration and understanding. As technology advances, the potential to harness energy and resources within this vast network may become feasible. This possibility presents both opportunities and ethical challenges, prompting reflection on our place within this cosmic mosaic and our responsibility to the universe. By pondering these questions, we not only deepen our cosmic understanding but also refine our role as caretakers of our planetary network, guiding us toward a future informed by the profound interconnectedness defining the universe.

The complex interplay of interconnected phenomena shapes our world, weaving a tapestry from the microscopic firing of neurons in our brains to the vast expanse of stellar systems in the cosmos. These patterns demonstrate a fundamental truth: when individual elements are linked together, they create something far greater than the mere sum of their parts. Neural networks exemplify the intricate nature of consciousness, while social interactions highlight the profound influence of human connection. The grand design of star clusters reflects the universal application of these principles. These interconnected systems remind us that small actions can initiate significant changes, encouraging us to recognize our influence within larger networks. By fostering mindful interactions, we can leverage these insights for societal advancement. As we move forward to explore how power laws shape various phenomena, we open the door to a deeper understanding of the unseen frameworks that govern our universe, inviting reflection on our role within this complex web.

Chapter Four
The Power Law Universe

In the quiet intersections of nature and society, an invisible force subtly influences our world: the power law. Picture yourself on the outskirts of a lively city, observing the intricate dance of people, vehicles, and structures. Despite the apparent chaos, a simple mathematical rhythm underlies this scene, guiding both the vast expanse of human civilization and the minute interactions of molecules. This chapter invites you to explore the mysteries of this profound pattern, revealing how it weaves together the very fabric of our universe.

Imagine being in a library, where a soft murmur of stories surrounds you. Each book contains a unique world, yet they all follow a curious rule—Zipf's Law—where word frequency decreases with rank. This principle extends beyond literature, evident in the layout of cities, where a few urban giants tower over numerous smaller towns. Such arrangements hint at the influence of power laws, which shape wealth distribution, natural phenomena, and the digital networks that connect us. This chapter will demystify how these patterns govern the economy, the environment, and the complex systems that underpin our interconnected world.

As we delve deeper, you will discover a universe sculpted by power laws, where the extraordinary emerges from the ordinary. These laws manifest in the unseen connections between diverse elements, from the synchronized flight of bird flocks to the sturdy design of the internet. Here, we examine the relationship between scale-free networks and their resilience, exploring the dynamic interactions that support both ecosystems and technological frameworks. This chapter will illuminate the profound impact of power laws, encouraging you to perceive the world through a lens of connected simplicity, where small changes can resonate across vast distances, inspiring us to apply these insights for transformative progress.

Zipf's Law in Language and Cities

Throughout history, the emergence of patterns in both nature and society has often surfaced in unexpected ways, offering insights into the hidden structures of our world. One such intriguing concept is Zipf's Law, a principle that may appear simple at first glance but holds significant implications across a range of fields. From the organization of words in a language to the design of urban spaces, Zipf's Law connects the abstract with the concrete, linking the complexities of human communication with city layouts. Imagine deciphering city dynamics through the frequency of word usage—a fascinating interplay of numbers and language, streets and architecture, all united by a mathematical principle. This exploration not only sheds light on existing systems but also provides guidance for shaping future developments.

As we delve into this topic, we uncover how Zipf's Law serves as a foundation for predictive modeling and urban planning, offering tools to anticipate growth patterns, allocate resources efficiently, and meet the needs of emerging smart cities. The law's influence extends further, enhancing artificial intelligence by improving the understanding of linguistic patterns in machine learning. As we probe deeper, the interconnections between language, urban environments, and intelligent systems become apparent, revealing an unexpected yet insightful tapestry of relationships. This journey through the universe of power laws not only deepens our understanding of these complex systems but also inspires innovative strategies to leverage these patterns for a more harmonious future.

Exploring the Mathematical Foundation of Zipf's Law

Zipf's Law is a captivating mathematical concept that highlights a unique pattern: the frequency of an event is inversely related to its rank. This intriguing distribution reveals order in systems that might initially seem chaotic. Emerging from statistical mechanics, Zipf's Law was first identified in the field of linguistics. George Zipf, an American linguist, noted that the most frequently used word in a language is about twice as common as the second most frequent, three times as common as the third, and this pattern continues. This insight underscores the inherent efficiency of communication, balancing the effort of the speaker with the comprehension of the listener.

Beyond linguistics, Zipf's Law is evident in urban hierarchies. Cities, akin to words in a language, follow a rank-size distribution. The largest city is typically twice as large as the second, three times as large as the third, and so forth. This consistency mirrors the self-organizing nature of urban development, influenced by socioeconomic factors and resource distribution. By examining urbanization through the lens of Zipf's Law, we can better understand growth dynamics, aiding in sustainable city planning and equitable resource allocation.

In predictive modeling, the implications of Zipf's Law are profound. Leveraging its principles, researchers and planners can craft models that predict linguistic trends or urban expansion with notable precision. In linguistics, it supports the creation of algorithms that enhance natural language processing, benefiting machine learning applications in translation and voice recognition. In urban scenarios, Zipf's Law offers a framework for projecting city growth, helping planners foresee infrastructure needs and address challenges like overpopulation and resource scarcity.

The potential applications of Zipf's Law extend into artificial intelligence and smart city development, where it fosters innovation. For AI, understanding this law refines algorithms that mimic human communication patterns, resulting in more intuitive interactions between humans and computers. In smart cities, Zipf's Law informs the design of systems that improve traffic flow, energy use, and public services, cultivating urban environments that are both intelligent and attuned to residents' needs. These applications highlight the critical role of mathematical insights in shaping future technologies.

Engaging deeply with Zipf's Law prompts reflection on its broader implications. Consider how this principle of proportionality might apply to other areas, such as wealth distribution or information spread in social networks. Can we harness these insights to create more equitable systems? Exploring these questions unlocks the potential of Zipf's Law, not only to comprehend the world around us but to actively shape it in ways that align with our aspirations for balance and harmony.

Comparative Analysis of Zipf's Law in Linguistic Structures and Urban Hierarchies

Zipf's Law, an intriguing mathematical concept, manifests in the patterns of language and city hierarchies. It highlights a similarity in the distribution of words in a language and the sizes of cities: both follow a pattern where a few are very common, while most are rare. In languages, this is seen in a small number of words being frequently used, with the rest appearing infrequently. Similarly, in urban environments, a few cities have large populations, while most are smaller. This parallel sheds light on the dynamics behind complex systems, sparking interest in the forces shaping such distributions across various fields.

Recent research has expanded our understanding of Zipf's Law, offering new insights for its use in predictive modeling. Linguistic patterns can enhance natural language processing algorithms, improving their ability to predict word usage in different contexts. In urban planning, recognizing the Zipfian distribution in city sizes can guide infrastructure development, resource distribution, and disaster preparedness. By recognizing these patterns, planners can better anticipate the demands of fast-growing urban areas, ensuring sustainable evolution and functionality.

The implications of Zipf's Law in artificial intelligence and smart cities are particularly exciting. AI systems that understand Zipfian distributions can optimize traffic flow, energy use, and emergency response in cities. The ability to anticipate and react to a city's changing needs in real-time could revolutionize urban management. As technology progresses, integrating Zipf's Law into AI-driven systems creates opportunities for innovation, allowing cities to function with remarkable efficiency and adaptability.

Exploring Zipf's Law in language and urban contexts also prompts philosophical reflections. It challenges us to ponder why unrelated systems exhibit similar patterns, encouraging exploration of universal principles governing complex systems. This inquiry invites interdisciplinary research, where linguists, urban planners, and data scientists collaborate to unravel the mysteries of pattern formation. By bridging these fields, we gain a deeper understanding of human behavior and societal development, fostering a holistic approach to global challenges.

Applying Zipf's Law in practical scenarios requires technical skill and an appreciation of its limitations and potential. Researchers and practitioners must avoid overfitting models to specific situations, ensuring adaptable and resilient solutions. Embracing both the predictability and unpredictability of complex systems allows us to harness Zipf's Law to create innovative solutions that resonate across scales, contributing to a more harmonious and sustainable world.

The Role of Zipf's Law in Predictive Modeling and Urban Planning

Zipf's Law is a compelling concept that finds application in both linguistics and urban planning, providing crucial insights for predictive modeling and city development. This principle, which explains the frequency of words in language and the hierarchical size of cities, uncovers a hidden order within seemingly chaotic systems. By comprehending this pattern, urban planners and data scientists can better predict and manage the complexities inherent in city growth and demographic changes. This understanding enables the development of predictive models that consider the irregular distribution of resources and population densities, offering a strategic edge in designing more efficient and resilient urban spaces.

In the field of predictive modeling, Zipf's Law offers a lens through which urban dynamics can be understood and forecasted. Recognizing that a small number of cities will host the majority of the population while numerous smaller towns persist allows planners to more accurately anticipate infrastructure and service needs. This foresight facilitates the more effective allocation of resources, ensuring that transportation systems, healthcare facilities, and educational institutions are aligned with actual population demands. Additionally, by integrating Zipfian distribution into models, planners can pinpoint areas likely to experience rapid growth or decline, enabling proactive responses to emerging trends.

The application of Zipf's Law also extends into smart cities and artificial intelligence, where its predictive power is used to enhance urban functionality. In smart city projects, data-driven decisions are crucial, and Zipfian models provide a solid foundation for AI algorithms that manage various aspects such as traffic and energy usage. By incorporating this principle, urban systems can be designed for self-regulation, effectively adjusting to changing demands. This adaptive capacity not only boosts efficiency but also promotes sustainability, aligning urban expansion with ecological considerations.

Recent advancements in computational power and data analytics have further amplified the utility of Zipf's Law in urban planning. Large-scale simulations and real-time data processing now enable more precise applications, tailoring strategies to the specific characteristics of individual cities. This customized approach allows planners to tackle issues like congestion or pollution with strategies that align with broader Zipfian trends. As urban areas continue to evolve, this detailed understanding will be essential in crafting solutions that are both innovative and empirically grounded.

Encouraging readers to consider the implications of Zipf's Law in their communities can inspire creative strategies for sustainable development. For instance, recognizing the influence of this principle might transform public policy or community initiatives. How could a deeper understanding of these patterns lead to a fairer distribution of resources or support more inclusive urban development? By exploring these questions, readers can apply these insights to drive meaningful change, contributing to the creation of cities that are not only efficient and functional but also vibrant and equitable.

Advanced Implications of Zipf's Law for Artificial Intelligence and Smart Cities

Zipf's Law is a captivating principle of statistical regularity that extends well beyond its typical application in linguistics and geography, opening new doors in fields like artificial intelligence and smart city development. In AI, this principle shapes natural language processing systems, enabling machines to interpret and produce human language with impressive accuracy. By understanding the predictable rank-frequency distribution of words, AI can better anticipate and adapt to human communication styles, enhancing its ability to engage in nuanced and context-aware interactions. This insight drives progress in AI by refining machine learning models, improving data compression, and optimizing search engines.

The influence of Zipf's Law also reaches into the growing domain of smart cities, where its insights can transform urban planning and administration. By acknowledging the consistent patterns in city sizes and structures, urban planners can design more efficient transportation systems, better allocate resources, and improve residents' quality of life. For example, the predictable distribution of city sizes can guide infrastructure development, ensuring fair resource distribution and preventing smaller cities from being overshadowed by larger ones. This approach helps create more balanced urban environments that cater to the varied needs of different communities.

As cities become more complex and interconnected, Zipf's Law gains significance in predictive modeling and decision-making. By applying this principle, city planners and policymakers can forecast urban growth, traffic congestion, and even the spread of information or diseases. Such predictive power is crucial for proactive management and crisis prevention. Understanding the distribution of resources and population densities can shape emergency response strategies, ensuring rapid and effective action when needed. This foresight is key to building resilience and adaptability, essential qualities for cities facing 21st-century challenges.

Beyond practical applications, Zipf's Law offers a lens to examine the philosophical and ethical aspects of AI and smart city technologies. By questioning how these systems interpret and apply such patterns, we can address concerns about privacy, equity, and representation. For instance, as AI models increasingly rely on language patterns, it is important to consider the potential reinforcement of biases present in the data. Similarly, smart city planners must consider the risk of unequal resource distribution to prevent technology from exacerbating existing inequalities. Engaging with these questions promotes a more thoughtful approach to technology integration, fostering systems that are both intelligent and compassionate.

The broader implications of Zipf's Law encourage us to rethink our relationship with technology and the world. By embracing the patterns underlying our languages and cities, we gain a deeper understanding of the interconnectedness of human systems and the natural environment. This holistic perspective fosters innovation and creativity, inspiring us to envision new possibilities for AI and smart cities that prioritize sustainability, inclusivity, and balance. As we continue to explore the vast potential of Zipf's Law, we are reminded of our shared responsibility to use this knowledge wisely, shaping a future that is both technologically advanced and deeply human.

Pareto Distributions in Wealth and Nature

Imagine a world where the allocation of resources, from the vibrant energy of urban environments to the tranquility of untouched wilderness, follows a mysterious yet consistent pattern. This pattern is governed by Pareto distributions, a concept often illustrated by the widely recognized 80/20 rule. This principle reveals that a small segment within any system exerts a substantial influence. In financial realms, this becomes evident as a minor group of individuals holds a significant share of global wealth. Yet, this pattern transcends human constructs, weaving its way through the natural world. In ecosystems, power laws manifest through the size distribution of species and the occurrence of natural events, hinting at simplicity underlying the complex tapestry of existence.

Our journey begins with unraveling the 80/20 rule's influence on economic systems, exploring how it shapes financial landscapes and informs decision-making. We then venture into the natural world, uncovering power laws in ecosystems and physical processes, offering a glimpse into nature's intricate choreography. As we delve deeper, the mathematical foundation of Pareto efficiency emerges, clarifying how these principles guide optimal resource allocation. Finally, we examine the dynamic interactions and scale invariance within ecosystems, revealing the mechanisms that sustain balance and resilience amidst change. Through this exploration, the Pareto distribution transforms from a mathematical notion to a powerful lens for understanding and impacting the world around us.

Understanding the 80/20 Rule in Economic Systems

The 80/20 rule, also known as the Pareto Principle, highlights the imbalance often present in economic systems, where a small number of causes lead to a large proportion of effects. In financial terms, this principle is evident when a minority holds the majority of wealth. Such wealth concentration reveals structural inequalities in economies, acting both as a symptom and a driver of systemic disparity. Recent research in behavioral economics indicates that these imbalances aren't solely due to historical factors or policy but are also influenced by inherent network dynamics that favor wealth accumulation among the affluent. This challenges conventional economic narratives and suggests policymakers focus on altering the fundamental properties of wealth distribution networks rather than relying solely on traditional redistribution methods.

The 80/20 rule is also prevalent beyond human economies, manifesting in nature's complex systems. Ecological observations frequently show power laws in action, where a few keystone species support most of an ecosystem's biodiversity. This dynamic highlights the resilience of natural systems, where critical components uphold the entire structure. Understanding these interactions allows conservation efforts to target these vital species, amplifying their positive impact on ecosystem stability. Advanced ecological studies employ network theory to map these connections, providing strategies for preserving biodiversity in changing environments and leveraging natural Pareto distributions for sustainable benefits.

Pareto efficiency adds another layer to the discussion. Central to welfare economics, it describes situations where resources can't be reallocated to improve one person's situation without worsening another's. This concept underscores the complexity of achieving optimal resource allocation, especially in economies where power laws skew distributions. Advances in computational economics have enabled simulations of Pareto efficiency scenarios, exploring pathways toward equitable distributions that maintain efficiency. These simulations challenge zero-sum perspectives, proposing innovative solutions that align equity with efficiency for a more inclusive economic framework.

The factors leading to Pareto distributions are rooted in complex systems theory, showing a hidden symmetry across various domains. The scale invariance seen in both ecosystems and economies suggests universal mechanisms at work. Complexity science suggests that these systems' self-organizing nature leads to Pareto-like distributions driven by feedback loops and emergent properties. This insight fosters interdisciplinary research, where ecological resilience insights can inform economic policy and vice versa. By adopting a holistic understanding, decision-makers can design interventions that utilize the self-regulating capabilities of complex systems, promoting stability and adaptability in both natural and human environments.

To leverage the potential of the 80/20 rule for positive change, individuals and organizations can identify the critical factors that disproportionately impact their goals. Entrepreneurs, for example, might allocate resources to high-impact projects or clients, maximizing returns with minimal investment. Similarly, environmentalists can focus conservation efforts on keystone species or ecosystems crucial for biodiversity. By adopting a strategic focus rather than spreading efforts too broadly, significant transformations can occur across various fields. Recognizing the power law dynamics empowers readers to drive meaningful change in personal endeavors and global challenges.

Natural Phenomena and the Emergence of Power Laws

In the natural world, power laws appear as universal principles governing various phenomena, unveiling order within what seems chaotic. These mathematical relationships highlight a small number of events causing the majority of an effect, such as the tallest trees in a forest capturing most sunlight or the largest earthquakes releasing most seismic energy. The frequent occurrence of power laws in nature emphasizes their role in shaping ecosystems' structure and dynamics, often determining resource distribution, species abundance, and interaction systems. This pattern of disproportionate influence challenges traditional linear models and encourages deeper exploration of the mechanisms behind such emergent behaviors.

A prime example of power laws in nature is the distribution of species within ecosystems. Biodiversity often exhibits a pattern where a few species dominate populations, while most remain less common, reflecting the well-known 80/20 rule. This distribution not only mirrors the competitive and cooperative dynamics among species but also highlights ecosystems' resilience and adaptability to environmental pressures. Ecological network research shows that these patterns provide stability, enabling ecosystems to endure disturbances and maintain functionality. Such insights advocate conservation strategies focusing on preserving keystone species and critical habitats, which significantly support ecological balance.

Recent studies have broadened our understanding of power laws by exploring their presence in geological phenomena. The frequency and magnitude of natural events like landslides, volcanic eruptions, and meteor impacts also adhere to power law distributions. These occurrences exhibit a fractal nature, showing self-similarity across scales. By analyzing these patterns, scientists can better predict the likelihood of catastrophic events, developing more effective risk management strategies. This application of power law analysis extends beyond prediction to enable proactive approaches to mitigate natural disaster impacts, safeguarding communities and preserving the environment.

In biological systems, power laws govern organisms' metabolic rates, offering insights into energy use efficiency across different life forms. The scaling laws describing how metabolic rates vary with body size reveal fundamental principles of life's organization, illustrating remarkable consistency across species. This understanding paves the way for advancements in fields like ecology, medicine, and bioengineering, where efficient energy use principles can inform sustainable technologies and medical interventions. Recognizing inherent efficiencies in natural systems can inspire human innovation, aligning technological development with ecological harmony.

As we uncover the pervasive nature of power laws, the challenge lies in translating these mathematical abstractions into actionable insights. By fostering an appreciation for natural phenomena's interconnectedness, we can develop strategies harnessing these patterns to address global challenges. This involves adopting a holistic perspective that values complexity and adaptability, encouraging us to think beyond simplistic cause-and-effect models. Thoughtful application of these principles can inspire solutions that are both innovative and sustainable, guiding humanity toward a future that respects and integrates the wisdom inherent in the natural world.

Mathematical Foundations of Pareto Efficiency

Pareto efficiency plays a crucial role in both mathematical economics and natural systems, highlighting the delicate balance between resource allocation and optimization. Originating from Vilfredo Pareto's insights into wealth distribution, this concept illustrates scenarios where reallocating resources cannot enhance one aspect without negatively impacting another. Its relevance spans various fields, from economic frameworks to ecological systems, showcasing its universal applicability. Pareto efficiency emphasizes the inherent trade-offs in complex systems, fostering a deeper understanding of how optimal states can be achieved and sustained in both human and natural environments.

Far from being a mere theoretical idea, Pareto efficiency serves as a practical analytical tool in economic systems. It offers a framework for assessing policy decisions, aiming for resource distributions that maximize societal benefits without unnecessary sacrifices. This principle is vital in welfare economics, where it shapes economic models to improve social welfare while recognizing limitations. It challenges policymakers to consider not only wealth distribution but also the underlying structures that either support or hinder equitable resource allocation. Embracing this efficiency helps societies develop economic strategies that balance prosperity with sustainability.

Beyond economics, Pareto efficiency finds intriguing parallels in nature. Ecosystems often reach equilibrium states where energy and resources are allocated to support diverse life forms without causing collapse. This balance mirrors the efficiency seen in economic systems, where changes can trigger cascading effects. Recent ecological studies reveal that energy transfer among trophic levels follows Pareto-like distributions, optimizing species survival and reproduction within resource limits. Understanding these principles provides valuable insights for conservation efforts, enabling strategies that maintain biodiversity while considering the intricate interdependencies of natural environments.

In recent years, Pareto efficiency has been explored in complex systems theory, offering new perspectives on scale invariance and self-organization. Researchers are discovering how Pareto principles appear in various networks, from internet node distributions to social system resilience. These insights drive interdisciplinary research, encouraging innovations that apply Pareto efficiency to challenges such as optimizing supply chains, enhancing network security, and improving urban planning. Recognizing the interconnectedness of these systems highlights the potential for Pareto efficiency to guide sustainable development, offering a roadmap for harnessing complexity to advance human progress.

As readers explore the mathematical foundations of Pareto efficiency, they are invited to consider its practical implications. By applying this concept, individuals and organizations can identify where minor adjustments yield significant benefits, whether in resource management, ecological conservation, or societal planning. This approach encourages continuous improvement, where pursuing Pareto-optimal states becomes a catalyst for innovation and growth. Engaging with these ideas invites a transformative perspective, recognizing the potential for small, thoughtful actions to drive meaningful change in the intricate web of global systems.

Complex Interactions and Scale Invariance in Ecosystems

In studying ecosystems, the notion of scale invariance unveils a complex web of interactions that challenge straightforward classification. This concept posits that certain patterns and processes remain consistent regardless of the observational scale. A clear example is found in species distribution within ecosystems, where the abundance often follows a power-law distribution: a few species are dominant while many remain rare. This pattern not only reflects ecological interactions but also mirrors economic disparities, like wealth distribution, suggesting the existence of deeper, universal principles.

Scale invariance in ecosystems extends beyond species distribution and includes the intricate interactions that maintain ecological balance. Food webs, for instance, display resilience due to their scale-free network structures, where some species act as highly connected hubs. This setup allows ecosystems to endure certain disturbances, as the removal of less connected species doesn't significantly disrupt the network. However, losing a hub species can trigger cascading effects, highlighting the delicate balance of these complex interactions. This understanding emphasizes the importance of conserving biodiversity, as protecting these hub species strengthens entire ecosystems against external pressures.

Recent research has revealed intriguing insights into how scale invariance appears in ecological dynamics over time. These studies show that ecosystems often exhibit fractal temporal patterns, where fluctuations in species populations or environmental variables display self-similarity across different time frames. These findings challenge traditional views of ecological stability, suggesting that ecosystems thrive on dynamic equilibrium characterized by continuous adaptation and transformation. This perspective advocates for conservation strategies that embrace natural variability and unpredictability, rather than imposing static conditions.

The interplay of complex interactions and scale invariance also invites examination of human influence on ecosystems. Human activities, such as urbanization and agriculture, introduce new variables that can either enhance or weaken natural power-law distributions. For instance, urban heat islands may alter local biodiversity, while farming practices can disrupt traditional predator-prey dynamics. Recognizing these influences encourages critical reflection on harmonizing human activities with natural processes, fostering coexistence that leverages the resilience of scale-invariant patterns.

As we reflect on scale invariance within ecosystems, several thought-provoking questions emerge. How can the principles of scale-free networks inform more sustainable agricultural systems? How might scale invariance enhance our understanding of climate change impacts on biodiversity? These considerations deepen our appreciation for the hidden order within ecological complexity and inspire innovative approaches to stewardship and sustainability. By aligning human endeavors with nature's intrinsic patterns, we can contribute to a harmonious balance that celebrates the diversity and complexity of life.

Scale-Free Networks in Complex Systems

Think back to a moment when you were astounded by the intricate connections that shape our world—how a single tweet can send ripples across continents or how a market shift can impact global economies. These aren't mere coincidences; they're deeply embedded in the network structures that define our modern lives. Here, we delve into the captivating world of scale-free networks, whose distinctive architecture supports a range of complex systems, from the neurons in our brains to the sprawling web of the internet. These systems boast a few highly connected hubs, crucial for maintaining the system's robustness and efficiency. However, this same architecture introduces vulnerabilities, as the failure of these hubs can trigger widespread repercussions.

As we explore scale-free networks, we unravel the principles behind their formation, such as preferential attachment, where connections build upon themselves. This journey will shed light on how these networks appear in real-world systems, transforming our understanding of how diverse entities, from biological ecosystems to social media platforms, function and evolve. Through this exploration, we'll uncover the power and potential of these networks, providing insights into their strengths and weaknesses and highlighting their applications across various fields. Each discovery invites us to appreciate the hidden order governing complex systems and empowers us to utilize these patterns for positive change.

Understanding the Architecture of Scale-Free Networks

Scale-free networks present a fascinating structure within complex systems, marked by a unique organization where a small number of nodes, known as "hubs," have a disproportionately large number of connections compared to the rest. This organization is not haphazard but follows a power-law distribution, indicating that the likelihood of a node having numerous connections decreases as the number of connections increases. This characteristic is evident in various systems, from the World Wide Web to biological networks, where connectivity is unevenly distributed. Such networks demonstrate resilience against random breakdowns but show vulnerability to deliberate attacks on the hubs. Grasping the complexity of scale-free networks is essential to appreciating their strengths and weaknesses, providing insights into both natural and engineered systems.

The architectural subtleties of scale-free networks showcase a compelling balance between order and disorder. Central to this configuration is the concept of preferential attachment, where new nodes tend to link to already well-connected nodes. This feedback mechanism resembles the social and economic principle of "the rich getting richer." Recent research has explored how preferential attachment impacts not just network growth but also its evolution over time. By studying real-world networks, like airline routes or social media platforms, scientists have gained understanding of how these systems maximize connectivity and efficiency while preserving some decentralization. The repercussions for understanding the spread of information or diseases and the diffusion of innovations through societies are significant.

Scale-free networks display an intriguing duality: they exemplify connectivity efficiency yet pose a security risk. The presence of highly connected hubs allows the network to function despite multiple node failures, but it becomes vulnerable if these critical hubs are targeted. This dual aspect has inspired innovative strategies to design more robust networks that retain the benefits of scale-free topology while reducing risks. For instance, incorporating redundant pathways or decentralized control systems can strengthen resilience. The ongoing challenge is to strike a balance between connectivity advantages and security necessities, requiring a continually evolving comprehension of network architecture.

In today's world, the study of scale-free networks has ventured beyond traditional boundaries into fields like neuroscience and quantum computing. For example, the scale-free arrangement of neural networks in the human brain is thought to aid efficient information processing and cognitive functions. Similarly, in quantum computing, deciphering the complex interactions within a scale-free framework may unlock new levels of computational power and speed. These advances highlight the importance of interdisciplinary approaches, merging insights from computer science, biology, and physics to unravel these networks' complexities. As research continues to progress, the potential for groundbreaking discoveries remains immense.

For those looking to apply scale-free network principles practically, the key is to leverage their strengths while addressing inherent weaknesses. One actionable strategy is to identify and reinforce critical hubs within a network, ensuring they withstand disruptions. Moreover, promoting diversity in connections rather than uniformity can lead to a more adaptable and robust network design. Encouraging innovation and cross-disciplinary collaboration can also foster new applications and solutions. By embracing the dynamic nature of scale-free networks, individuals and organizations can optimize their systems for enhanced efficiency and resilience, paving the way for transformative change in an interconnected world.

The Role of Hubs in Network Resilience and Vulnerability

In the complex web of scale-free networks, central nodes known as hubs hold significant sway over the network's stability and weaknesses. These hubs, with their multitude of connections, act as crucial links, offering both strength and potential vulnerabilities. Their existence allows most nodes to stay linked through just a few steps, a phenomenon often termed the "small-world" effect. This design confers resilience on the network, enabling it to endure random failures gracefully. The removal of a single node, even a hub, generally does not sever the overall connectivity. However, the network's heavy reliance on hubs introduces a certain vulnerability. If a hub is deliberately targeted or fails due to external pressures, the repercussions can ripple throughout the system, causing significant fragmentation or even collapse.

The dual role of hubs in these networks invites exploration of their impact across various fields. In technology, the Internet serves as a prime example, where major servers and routers function as hubs facilitating global communication. Their robustness is a testament to thoughtful planning, managing vast traffic with minimal delay. Yet, the 2016 Dyn cyberattack exposed weaknesses, as the temporary disablement of critical hubs disrupted access to numerous websites. This incident highlights the need for strategic planning and redundancy to minimize the effects of potential hub failures, illustrating a delicate balance between efficiency and security.

In social networks, hubs are individuals with extensive connections who often act as influential figures. These social hubs are instrumental in spreading information, ideas, and trends, shaping public opinion and social movements. They can amplify messages and mobilize communities effectively. However, this influence can also facilitate the rapid spread of misinformation, posing societal challenges. By understanding the dynamics of hubs in social networks, stakeholders can devise strategies to harness positive influence while curbing harmful content.

In biological systems, the human brain features hubs known as cortical hubs, which enable efficient communication across different brain regions. These hubs are vital for cognitive processes, allowing quick information processing and decision-making. Recent neuroscience research emphasizes the importance of maintaining the health and functionality of these hubs to prevent cognitive decline. Insights from studying these neural networks offer promise for developing interventions targeting specific hubs to enhance cognitive resilience or treat neurological disorders.

To harness the power of hubs in scale-free networks, individuals and organizations can take proactive measures. Incorporating redundancy into critical infrastructures, diversifying connections, and fostering decentralized systems can bolster resilience. In social contexts, promoting diverse voices and nurturing a culture of critical thinking can mitigate the risks of misinformation. By applying these strategies, stakeholders can navigate the complexities of scale-free networks, leveraging their strengths while protecting against vulnerabilities. This nuanced understanding empowers readers to engage thoughtfully with complex systems, contributing to a more resilient and interconnected world.

Emergence of Scale-Free Networks Through Preferential Attachment

Scale-free networks arise from the intriguing process of preferential attachment, which highlights the dynamic evolution of these complex systems. This principle describes how networks expand by favoring already well-connected nodes, leading to a few "hubs" with numerous links and many smaller nodes with fewer connections. Preferential attachment is not merely theoretical; it is evident in various real-world scenarios, from the internet's framework to social networks and biological systems like brain neuron connections. Starting with a few nodes, as new ones join, they gravitate towards the more prominent nodes, reinforcing the network's unique structure. This self-organizing nature enhances the network's adaptability and robustness, though it also introduces specific vulnerabilities.

A vivid illustration of preferential attachment is seen in the development of the World Wide Web. As new websites emerge, they often link to popular sites with high visibility, like Wikipedia or Google. Such linking behavior aligns with preferential attachment, crafting a network dominated by a few highly connected hubs that streamline information flow and accessibility. This structure enables the internet's efficient operation, ensuring information is readily available and widely spread. However, these hubs also represent critical points of failure; their disruption could have far-reaching effects. This balance of stability and fragility prompts ongoing research into optimizing network resilience while preserving the advantages of a scale-free design.

The phenomenon of scale-free networks transcends digital or technological boundaries, appearing widely in biological and social contexts. In cellular frameworks, proteins interact in scale-free patterns, where a few proteins engage with many others, influencing various cellular functions. This architecture is crucial for comprehending cellular responses to environmental changes and treatments. Social networks, similarly, exhibit preferential attachment, where individuals with extensive social connections tend to attract more connections, magnifying their influence. Exploring these patterns across diverse fields underscores the universality of preferential attachment in shaping intricate networks.

Recent studies have explored the subtleties of preferential attachment, investigating how changes in initial conditions, growth rates, and node characteristics affect network structure. Research indicates that introducing randomness or external influences into the attachment process can create varied network topologies, each offering distinct properties and applications. This knowledge is vital for designing artificial networks, such as transportation or communication systems, where understanding and manipulating preferential attachment can enhance efficiency and resilience. By identifying the factors driving network growth, researchers and practitioners can better anticipate and manage complex system development.

Grasping the mechanisms behind preferential attachment provides practical insights for leveraging or disrupting scale-free networks. For example, in public health, understanding how diseases propagate through social and transportation networks can guide strategies to control outbreaks by targeting highly connected nodes. In business, companies can harness preferential attachment by strategically positioning themselves within key industry hubs, boosting their influence and growth. By applying these principles, individuals can engage proactively with the dynamics of scale-free networks, whether improving organizational structures, enhancing technological systems, or fostering social change, paving the way for innovation and resilience in a rapidly evolving world.

Applications of Scale-Free Network Theory in Real-World Complex Systems

In the intricate world of complex systems, scale-free networks stand out for their unique architecture, where a few nodes, or hubs, have numerous connections, while many other nodes are less connected. This distinctive design is not just theoretical; it forms the backbone of many real-world systems. From the sprawling web of the internet to the interconnected neurons in the human brain, scale-free networks exemplify nature's inclination towards efficiency and resilience. These systems typically follow a power-law distribution, meaning a node's likelihood of having a particular number of connections decreases sharply as the number of connections increases. This characteristic enables them to grow and adapt, meeting new demands and challenges over time.

A notable feature of scale-free networks is their robustness. Hubs provide these networks with significant durability against random disruptions. For example, in biological systems, the presence of highly connected nodes ensures functionality even when some components fail. Similarly, in telecommunications, this robustness results in fewer service interruptions despite random outages. Nevertheless, this strength can also pose a weakness. While resilient to random failures, scale-free networks are susceptible to targeted attacks on their hubs. Understanding this duality is crucial for creating systems that are both efficient and secure, driving continuous research into protective measures for critical nodes.

The formation of scale-free networks often involves preferential attachment, where new nodes tend to link to more connected nodes. This self-reinforcing mechanism is evident in many fields, from the growth of social media to the development of scientific citation networks. As a node accumulates more connections, it becomes increasingly attractive to new nodes, fostering a dynamic where the rich get richer. Recent research has delved into this growth model, revealing how minor changes in attachment rules can lead to vastly different network structures. For instance, introducing randomness or external constraints can alter the balance, resulting in networks that deviate from the typical scale-free pattern.

Beyond theoretical exploration, the principles of scale-free networks have driven innovation across various fields. In epidemiology, understanding human interaction patterns has informed strategies to control disease spread. By targeting hubs—individuals or locations with high connectivity—health interventions can be more precisely directed, enhancing their effectiveness. In urban planning, insights from scale-free networks aid in developing sustainable infrastructures by optimizing traffic flow and resource distribution, ensuring cities remain functional as they grow. These practical applications demonstrate the versatility and importance of scale-free network theory in tackling contemporary challenges.

As we explore the complexities of scale-free networks, several intriguing questions emerge. How can we leverage the strengths of these networks while addressing their vulnerabilities? What lessons can we draw from natural systems to enhance artificial networks? Answers to these questions could transform fields from cybersecurity to ecological conservation. By embracing the principles of scale-free networks, individuals and organizations can create systems that are not only more efficient but also more adaptable to the unpredictable dynamics of today's world. Readers are encouraged to consider these possibilities, envisioning how the power of connectivity can drive positive change in their own areas of influence.

Power laws act as the unseen forces shaping our world, influencing how resources are distributed, how societies are organized, and the very architecture of the complex systems that surround us. From the linguistic elegance of the Zipfian principle to the widespread Pareto distributions in economics and ecology, these concepts emphasize the significant impact a small number can have over the majority. By exploring scale-free network structures, we uncover insights into both the resilience and weaknesses found in natural and human-designed systems. Recognizing these dynamics equips us to anticipate changes and strategize for fairer resource allocation and sustainable development. As we ponder the profound effects of power laws, we face a crucial question: how can we use this knowledge to create a world where connectivity and abundance benefit everyone more equally? This understanding sets the stage for our next exploration, where we will uncover how synchronization and rhythm unlock further mysteries of the universe.

Chapter Five
Synchronization Patterns

In the vibrant core of a city that never sleeps, where thousands of lives intersect in a chaotic yet harmonious dance, an extraordinary event unfolds each evening. At exactly 6 PM, the traffic signals along the main thoroughfare align with remarkable precision, ushering a smooth cascade of vehicles that defies the typical rush-hour turmoil. This coordination resembles the hidden symmetries found in nature and the universe. It is not just a feat of engineering genius; it reflects a deeper, universal order governing systems far beyond our bustling streets.

This invisible thread of alignment stretches across the cosmos, linking the rhythmic heartbeat of living creatures to the intricate patterns of financial markets. Within this complex tapestry, quantum particles display mysterious connections that transcend space and time. Meanwhile, our bodies echo the cycles of the natural world, from the steady rhythm of our hearts to the synchronized glow of fireflies on warm summer nights. These patterns demonstrate how entities—be they particles or people—align their actions to create phenomena that surpass the sum of their individual parts.

As we delve into these fascinating expressions of alignment, we uncover the mathematical truths that bind them. Each example offers a glimpse into the elegant simplicity underlying complex systems, inviting us to explore how these principles can foster harmony and resilience in our interconnected world. By grasping the power of coordination, we gain insight into how small changes can ripple through vast networks, sparking global transformations. Through this perspective, the chapter weaves a narrative that is both enlightening and empowering, encouraging readers to explore the potential of alignment in their own lives and communities.

Quantum Entanglement

Imagine the intricate dance of subatomic particles, a performance that defies our usual grasp of space and separation. Quantum entanglement, sometimes called "spooky action at a distance," poses a challenge to our perception of reality. In this curious phenomenon, particles become so closely linked that a change in one instantaneously affects the other, regardless of distance. This chapter invites you to delve into the complexities of these connections, exploring how they form and persist, and what they mean for both the small and large scales of our universe.

What makes this topic fascinating is not just its theoretical intrigue but also its practical impact across various fields. As we explore further, you'll see how entanglement plays a crucial role in advancing quantum computing and cryptography, hinting at a future where information is handled in ways once thought impossible. We'll examine the experimental techniques used to observe this elusive phenomenon, showcasing the precision and creativity needed to capture these moments of connection. Finally, we'll ponder the broader implications of entanglement for our understanding of reality, challenging us to reconsider what we know about the universe. This exploration serves as a bridge, connecting the ethereal realm of quantum physics with tangible changes across diverse areas, setting the stage for a multifaceted journey into the patterns of synchronization that lie ahead.

The Phenomenon of Non-Local Correlations

In the world of quantum mechanics, non-local correlations, especially seen in quantum entanglement, push the boundaries of classical physics. Quantum entanglement emerges when particles become interconnected in a way that the state of one cannot be described independently from the others, regardless of the distance between them. This concept challenges the conventional idea that objects are only influenced by their immediate environment, offering a striking contradiction to traditional physical laws. The exploration of non-local correlations not only enriches theoretical physics but also provides profound insights into the universe's interconnectedness, suggesting that our grasp of reality remains incomplete.

Recent breakthroughs in quantum physics have deepened our understanding of these non-local phenomena. Experiments have shown that entangled particles can instantaneously affect each other's states over great distances. One notable experiment with entangled photons demonstrated that altering the polarization of one photon instantly impacted its distant counterpart, even when kilometers apart. These observations not only reinforce the concept of quantum entanglement but also highlight its potential to transform technology. Non-locality is sparking innovative ideas about information transfer, paving the way for advancements in areas like quantum computing and secure communication networks.

The mysterious nature of quantum entanglement urges a reevaluation of causality and locality. Traditional cause-and-effect principles are challenged as entangled particles seem to interact at speeds surpassing light. This leads to engaging discussions on the fundamental principles of physics and the essence of reality. Some physicists suggest that the universe is fundamentally interconnected beyond space and time, while others consider our perception of separateness merely an illusion. These debates drive scientific inquiry and invite philosophical reflection on existence itself.

Beyond its theoretical allure, the practical applications of non-local correlations are vast and transformative. In the nascent field of quantum computing, entanglement is essential for developing qubits, which can process data at unprecedented speeds. This advancement could revolutionize complex problem-solving tasks, from cryptography to optimization, by enabling computations currently impossible with classical computers. Additionally, quantum entanglement holds promise for enhancing communication security through quantum cryptography, offering encryption methods that are theoretically unbreakable due to the unique properties of entangled particles. These technological innovations promise to reshape industries and redefine the limits of what we can achieve.

As we continue to explore the mysteries of non-local correlations, we are encouraged to consider their broader implications for our understanding of the cosmos. Investigating quantum entanglement not only pushes the boundaries of scientific knowledge but also inspires us to envision a world where connectivity transcends physical limits. By embracing the potential of these quantum phenomena, we are advancing technology while expanding our comprehension of the universe and our place within it. This journey of discovery, fueled by curiosity and innovation, holds the promise of unlocking new dimensions of reality and inspiring future generations to explore the uncharted territories of science and philosophy.

Entanglement in Quantum Computing and Cryptography

Quantum entanglement, a captivating intersection of theoretical physics and applied technology, serves as the foundation for innovations in quantum computing and cryptography. This phenomenon binds particles in such a way that the state of one affects another, irrespective of the distance between them. This instantaneous link forms the core of quantum computing, where qubits replace traditional bits. Due to entanglement, qubits can exist in multiple states at once, allowing quantum computers to solve complex problems at unprecedented speeds. This leap in computational capability has the potential to transform industries like healthcare, finance, and artificial intelligence, akin to the shift from candlelight to electricity, unveiling new possibilities in previously uncharted territories.

In cryptography, entanglement ushers in a new era of security. Quantum key distribution (QKD) uses the properties of entanglement to establish secure communication channels. Unlike classical encryption, QKD employs entangled particles to generate keys, making any eavesdropping attempts immediately noticeable. This groundbreaking method could render many traditional cryptographic techniques outdated, protecting sensitive data in an increasingly connected world. As cyber threats become more sophisticated, the robust encryption provided by quantum cryptography is not just advantageous but necessary, offering a future where data integrity and confidentiality are secure from malicious attacks.

Recent advancements in experimental methods have propelled the practical application of quantum entanglement to unprecedented levels. Global laboratories are pushing boundaries with entangled states to improve quantum coherence and error correction—vital steps in achieving fully operational quantum computers. These scientific pursuits are supported by growing investments from both private sectors and governments, who recognize the strategic significance of quantum technology. Collaborations between academia and industry create a dynamic environment where theoretical insights quickly translate into practical technological advances. The interplay between research and development is like a dance, where each advance in understanding entanglement accelerates its implementation in real-world applications.

The impact of entanglement goes beyond technical achievements, prompting profound philosophical inquiries about the nature of reality. It challenges conventional intuitions, suggesting a universe where locality and determinism are not absolute. This invites a reevaluation of fundamental concepts such as causality and connectivity, urging a reconsideration of the cosmos's fabric. By engaging with these questions, scientists and philosophers enter a dialogue that transcends disciplines, offering a richer understanding that weaves science with human experience. Such explorations highlight the dual nature of entanglement as both a technological frontier and a philosophical catalyst.

For those captivated by the potential of entanglement in quantum domains, there are practical ways to engage with this emerging field. Aspiring enthusiasts can explore educational resources like online courses and workshops that simplify quantum mechanics and its applications. Participating in hackathons or joining research groups focused on quantum technologies provides hands-on experience and exposure to cutting-edge developments. By fostering curiosity and a willingness to learn, individuals can contribute to the quantum revolution, whether through direct participation or by supporting its societal implications. The journey into the quantum realm is more than a professional pursuit; it's a voyage of discovery inviting us all to reflect on our place in the broader cosmic narrative.

Experimental Techniques for Observing Quantum Entanglement

Quantum entanglement, a fascinating phenomenon in physics, can be observed using advanced experimental techniques that link theoretical predictions with real-world observations. Central to these techniques is the precise measurement of correlations that emerge when particles become entangled. Defying classical logic, quantum entanglement showcases connections that remain intact regardless of distance. Researchers leverage this intriguing characteristic by employing sophisticated experiments like Bell tests to confirm entanglement. These experiments often involve generating pairs of entangled photons and assessing their polarization states, revealing correlations that surpass classical explanations. The careful design of these experiments ensures that no local hidden variables can explain the outcomes, providing compelling evidence of quantum entanglement's unique nature.

Recent technological advancements have transformed the methods for generating and observing entangled states, expanding the possibilities beyond previous expectations. High-precision lasers and advanced detectors are crucial in these experiments, enabling scientists to maintain and detect entangled states with remarkable accuracy. Platforms such as quantum dots, trapped ions, and superconducting circuits offer novel approaches to studying entanglement, each with distinct advantages tailored to specific research goals. By combining these cutting-edge tools with refined statistical techniques, researchers can filter out noise and identify genuine quantum correlations, deepening our understanding of the quantum world. These breakthroughs not only enhance our comprehension of entanglement but also pave the way for practical applications in emerging technologies.

The application of these experimental methods extends beyond observation, influencing fields like quantum computing and cryptography. Entanglement forms the foundation for quantum information processes, where interconnected qubits enable parallel computations and secure communication. By refining techniques to manipulate entangled states, scientists unlock new possibilities for quantum technologies that promise to outperform classical systems in certain tasks. For example, quantum key distribution uses entangled particles to create encryption keys that are theoretically unbreakable, protecting data in our increasingly digital world. The ongoing enhancement of entanglement observation techniques directly contributes to these innovative applications, highlighting the synergy between fundamental research and practical advancement.

Despite significant progress, observing quantum entanglement remains challenging, filled with complexities that spark debates and inspire diverse methodologies. Different interpretations of quantum mechanics offer various explanations for entanglement's nature, each with unique experimental implications. The debate between supporters of the Copenhagen interpretation and advocates of many-worlds or pilot-wave theories exemplifies this diversity, encouraging broader exploration of experimental approaches. Researchers continually develop new methods to test these interpretations, aiming not only to observe entanglement but also to deepen our understanding of its implications for reality. This dynamic landscape fosters a rich environment for scientific inquiry, where each experiment enhances our nuanced understanding of the quantum realm.

As we refine techniques for observing quantum entanglement, the potential for groundbreaking discoveries remains vast. Researchers are investigating ways to extend entanglement to larger systems, bridging the micro and macroscopic worlds in unprecedented ways. These efforts could revolutionize our understanding of interconnectedness in the universe, challenging conventional notions of space, time, and causality. As we explore the quantum tapestry, thought-provoking questions arise: Can entanglement connect distant galaxies, or might it reveal hidden dimensions of reality? These inquiries fuel scientific imagination and inspire practical steps to harness quantum phenomena for humanity's benefit. By embracing a collaborative and open-minded approach, the scientific community continues to push the boundaries of possibility, driven by the promise of unlocking new dimensions of knowledge and innovation.

Implications of Entanglement for Understanding Reality

Quantum entanglement offers profound insights that challenge traditional beliefs about separateness and locality, suggesting a universe far more interconnected than previously conceived. This phenomenon, where particles remain linked regardless of distance, introduces the intriguing concept of non-locality. Once an abstract mystery, non-locality now offers a glimpse into the universe's fabric, hinting at a deeply unified cosmos where actions in one location can instantaneously affect distant counterparts. These revelations prompt a reevaluation of classical physics paradigms and invite exploration of a reality where interconnectedness transcends conventional boundaries.

In the fields of quantum computing and cryptography, entanglement moves beyond theoretical curiosity, transforming into a practical tool poised to revolutionize technology. Entangled particles facilitate quantum superposition and parallelism, laying the groundwork for unprecedented computational power. Quantum cryptography harnesses these principles to establish unbreakable communication channels, where any eavesdropping attempt disrupts the entangled state, ensuring ultimate security. These applications highlight entanglement's transformative potential, offering a vision of a future where information processing and security are fundamentally redefined.

Advancements in experimental techniques have significantly enhanced our ability to observe quantum entanglement, allowing scientists to delve deeper into the quantum realm. Sophisticated experiments with entangled photons and ions have confirmed entanglement's counterintuitive properties, aligning with quantum mechanics predictions. Techniques like Bell test experiments have addressed loopholes, reinforcing the reality of entanglement beyond theoretical speculation. These scientific endeavors not only cement entanglement as a cornerstone of modern physics but also inspire innovative methodologies for exploring the universe's most enigmatic phenomena.

Beyond the scientific community, entanglement's implications resonate with philosophical perspectives on reality and consciousness. The interconnectedness suggested by quantum entanglement echoes ancient philosophical traditions emphasizing unity and oneness. In contemporary discourse, entanglement prompts reconsideration of consciousness and its potential link to quantum processes. This intersection of science and philosophy nurtures a holistic view of existence, encouraging dialogues that bridge empirical inquiry and existential reflection, challenging individuals to expand their understanding of reality.

As we stand on the brink of quantum exploration, practical insights from entanglement urge us to apply these principles beyond theoretical and technological realms. Embracing the interconnectedness inherent in entanglement can inspire collaborative approaches to global challenges, fostering a mindset of unity and shared responsibility. This perspective nurtures the potential for innovative solutions, fostering a world where the lessons of quantum science inform social, environmental, and technological advancements. By contemplating the profound implications of entanglement, we are invited to rethink our place within the cosmos, exploring the boundless possibilities that arise when we acknowledge the fundamental interconnectedness of all things.

Biological Rhythms

Biological rhythms form a fascinating and essential part of life, guiding the unseen choreography of existence within every living being. These cycles, which span from daily to seasonal, are not mere scientific curiosities; they are the very clocks that regulate life's essential functions. From the predictable rise and set of the sun to the intricate cellular activities within, these cycles create the foundation upon which life flourishes. As we delve into these natural patterns, we reveal their significant roles in shaping behavior, controlling physiological processes, and influencing entire ecosystems.

These rhythms are diverse and complex, encompassing various cycles with different durations and effects. The 24-hour circadian rhythms, for example, are crucial in aligning our sleep patterns, hormone release, and mental alertness with the Earth's rotation. Beyond these daily cycles, organisms also experience shorter ultradian and longer infradian rhythms, each impacting life in unique ways. The process of entrainment, where external signals synchronize these internal clocks, highlights the deep connection between living organisms and their environments. On a molecular scale, the forces driving these cycles reveal a complex interplay of cellular activities and genetic expressions. By exploring each aspect, we gain insights into how these patterns influence life, offering a deeper understanding of the interconnected nature of biological systems.

Circadian Rhythms and Their Influence on Organism Behavior

Circadian rhythms are the innate 24-hour cycles that guide both physiological and behavioral processes, crucially linking organisms with their surroundings. These internally generated cycles primarily align with external environments through light cues, allowing organisms to predict and adjust to daily changes. In mammals, the suprachiasmatic nucleus (SCN) in the hypothalamus acts as the main circadian clock. Recent studies reveal its intricate interaction with peripheral clocks located throughout the body, each playing a role in maintaining temporal order. This coordination is vital for efficient energy use, sleep regulation, and metabolic functions, highlighting the significant impact circadian rhythms have on behavior.

The circadian system's complexity goes beyond adapting to light-dark cycles, influencing a broad range of physiological activities. Hormones like cortisol and melatonin follow distinct circadian patterns, affecting alertness and sleepiness. Recent breakthroughs have identified molecular elements, such as CLOCK and BMAL1 proteins, that regulate these rhythms at the genetic level. These findings have propelled the field of chronobiology forward, revealing how disruptions in circadian rhythms can lead to health issues such as metabolic disorders, depression, and even cancer. Understanding these molecular complexities opens up new therapeutic possibilities, from gene editing to drugs that aim to restore circadian balance.

The role of circadian rhythms in adaptive evolution is gaining attention. Different species have developed varied circadian strategies to enhance survival in their unique habitats. Nocturnal animals, for example, have evolved keen senses and metabolic traits suited for low-light living, demonstrating how circadian timing variations drive evolutionary success. These insights challenge traditional views and encourage scientists to reconsider the influence of circadian rhythms on ecological interactions and biodiversity. As research progresses, questions arise: How might climate change affect wildlife's circadian alignment? Could adjusting circadian patterns offer new conservation methods? These inquiries pave the way for novel exploration and practical application.

Understanding circadian rhythms provides actionable insights for improving human health and productivity. Aligning daily activities with circadian phases can boost cognitive performance, optimize work schedules, and enhance well-being. For instance, exposure to natural light during the day and reducing artificial light at night significantly improves sleep quality. Additionally, the growing field of chronotherapy seeks to apply circadian principles to optimize medical treatment timing, increasing effectiveness and reducing side effects. This approach highlights the potential for circadian research to revolutionize healthcare, making it more personalized and effective.

Exploring the circadian dimension invites reflection on the broader implications of time in biological systems. As science continues to unravel these rhythms' complexities, one might ponder the philosophical questions they pose: How do our temporal patterns shape our world perception? Can understanding and respecting these natural cycles foster a more harmonious existence? By considering these questions, readers may reflect on how aligning with our internal clocks could contribute to personal growth and societal advancement. The study of circadian rhythms thus mirrors the interconnectedness and cyclical nature of life, inspiring a deeper appreciation for the delicate balance that sustains us.

Ultradian and Infradian Cycles in Biological Systems

Ultradian and infradian cycles, while not as widely recognized as the circadian rhythm, are crucial for understanding the complex timing within biological systems. Ultradian cycles, which occur more frequently than once every 24 hours, significantly influence processes such as hormone release, REM sleep phases, and cellular metabolism. These shorter cycles manifest in the daily fluctuations of alertness and energy, impacting productivity and cognitive function. By identifying these patterns, both individuals and organizations can better optimize work schedules and activities, aligning tasks with natural peaks in efficiency and focus.

In contrast, infradian cycles span longer than a single day, encompassing phenomena like menstrual cycles in mammals and seasonal behavior in flora and fauna. Recent research has underscored the impact of infradian rhythms on immune function and mental health, suggesting a profound connection to overall well-being. An understanding of the hormonal changes associated with these rhythms enables more personalized healthcare approaches, enhancing the management of conditions such as seasonal affective disorder and premenstrual syndrome.

The alignment of these biological cycles with environmental cues is vital for maintaining internal balance. The process of entrainment, where external factors like light and temperature synchronize internal clocks, ensures these cycles remain in harmony with external conditions. Advances in chronobiology have shown how disruptions in this alignment, often caused by the demands of modern life and artificial lighting, can lead to negative health effects. Addressing these disruptions through lifestyle changes, like strategic exposure to natural light and careful regulation of sleep patterns, can improve alignment with these innate cycles.

Molecular insights into ultradian and infradian rhythms have transformed our understanding of their mechanisms. Researchers have identified genetic and molecular pathways that control these cycles, highlighting the proteins and genes that rhythmically oscillate. This knowledge has significant implications for biotechnology and medicine, paving the way for innovations in drug delivery systems that are timed with these rhythms, thereby enhancing their effectiveness and reducing side effects.

As our understanding of ultradian and infradian cycles deepens, we are encouraged to rethink how we structure our lives and environments. Acknowledging these rhythms encourages the adoption of more rhythm-aware practices in sectors such as education, healthcare, and the workplace. By embracing these natural cycles, we can create environments that honor biological timing, fostering a more harmonious interaction between our internal clocks and the surrounding world.

Role of Entrainment in Synchronizing Biological Clocks

Entrainment is like the conductor in the grand symphony of life, guiding the synchronization of internal clocks within organisms. This process, where external signals align internal timekeepers, is crucial for organisms to adapt and flourish in their dynamic environments. Factors such as light, temperature changes, and social interactions act as synchronizing cues, or zeitgebers, that adjust internal rhythms to match the outside world. This alignment ensures that processes like sleep cycles, hormone regulation, and metabolism occur at optimal times, enhancing health and functionality.

Recent breakthroughs in chronobiology have revealed intriguing details about the molecular mechanisms of entrainment. At the cellular level, proteins like cryptochromes and clock genes such as PER and TIM cycle in a feedback loop, regulating circadian rhythms. These molecular timekeepers respond to environmental cues, triggering cascades that adjust the organism's internal clock. For example, exposure to morning light initiates a series of reactions in retinal cells, which then influence the suprachiasmatic nucleus—a small brain region—ultimately resetting the master clock. This complex interaction highlights the elegance of evolutionary design, with organisms finely tuned to their environments through the intricate system of entrainment.

The effects of entrainment go beyond physiological processes, impacting behavior and well-being significantly. Misalignment, such as that caused by jet lag or shift work, can disrupt biological clocks, adversely affecting mood, cognition, and health. Understanding entrainment principles enables interventions to counteract these effects. Light therapy, for instance, is used to realign circadian rhythms in those with seasonal affective disorder or sleep disorders. By exposing individuals to specific light wavelengths, it is possible to reset their internal clocks, improving mood and cognitive function.

On a larger scale, entrainment's role in harmonizing biological clocks illustrates the interconnectedness of life on Earth. Just as individual rhythms align within organisms, ecosystems display synchronized patterns, with predator-prey cycles and plant blooming times responding to environmental triggers. This synchronization fosters resilience and stability in ecological communities, ensuring survival and resource optimization. Entrainment, therefore, aids adaptation at the organism level and contributes to the dynamic balance of ecosystems, emphasizing a universal pattern of connection.

Considering the transformative potential of entrainment, one might contemplate how these principles could enhance human systems. Could societal structures benefit from a deeper understanding of biological rhythms? Might work and educational systems be redesigned to align with natural cycles, boosting productivity and well-being? These questions invite exploration into how biological synchronization could inform human endeavors, promoting a balanced harmony between technological progress and natural rhythms. Embracing the lessons of entrainment offers a path to fostering resilience and sustainability in both individual lives and collective systems.

Molecular Mechanisms Underlying Rhythmic Patterns in Cells

The intricate choreography of cellular rhythmic patterns is a testament to the precision and complexity inherent in biological life. Central to this dance are molecular oscillators, which serve as the timekeepers vital for orchestrating cellular processes. These oscillators operate through feedback loops composed of genes and proteins, ensuring that cellular functions proceed with impeccable timing. At the core of this mechanism is the transcription-translation feedback loop, where genes are cyclically activated and deactivated, regulated by the concentration and presence of specific proteins. This sophisticated interaction is not just a biochemical curiosity but essential for maintaining the physiological balance within organisms.

Recent strides in molecular biology have illuminated the crucial role of clock genes in managing these cellular rhythms. Research highlights genes like PER and CLOCK, which form an intricate network regulating gene expression timing. This governance extends to numerous biological processes, from cell division to metabolic pathways. Advanced techniques like CRISPR and real-time imaging have demonstrated how even slight disruptions in these genes can lead to significant physiological repercussions, emphasizing the precision needed for proper rhythmic functioning. Such insights pave the way for understanding how cellular rhythms influence health and disease, offering potential therapeutic targets for conditions arising from circadian misalignment.

Exploring cellular rhythms uncovers the intriguing concept of entrainment, where external cues, known as zeitgebers, align internal clocks with the external world. Factors like light, temperature, and nutrient availability play pivotal roles in synchronizing cellular rhythms. The molecular mechanisms facilitating entrainment involve elaborate signaling pathways that convert external stimuli into internal biochemical signals. For example, light exposure initiates a series of reactions that adjust the molecular clock's timing to align with day-night cycles. This synchronization is crucial for organisms to adapt to environmental changes, highlighting the evolutionary significance of rhythmic patterns.

Rhythmic patterns extend beyond circadian rhythms to include ultradian and infradian cycles, which operate on shorter or longer timescales, respectively. These cycles are vital for various physiological processes, such as hormone secretion and cellular repair. The molecular basis of these rhythms involves unique sets of genes and proteins that work in concert with the core circadian machinery. Understanding these cycles offers profound insights into the temporal organization of cellular functions, showcasing the multilayered complexity of biological rhythms. Researchers are continually uncovering how these rhythms interact, revealing a complex tapestry of temporal regulation that is both intricate and essential.

As molecular biology advances, new technologies and interdisciplinary approaches promise to unravel the mysteries of cellular rhythms further. By integrating systems biology, computational modeling, and advanced imaging techniques, scientists are poised to discover how cellular rhythms govern the dynamic interaction between genes, proteins, and environmental factors. This expanding knowledge not only deepens our understanding of fundamental biological processes but also empowers us to leverage these rhythms for innovative applications in medicine, agriculture, and beyond. By appreciating the nuanced choreography of molecular mechanisms underlying rhythmic patterns, we gain the tools to foster resilience and adaptability in the face of ever-evolving challenges.

Market Correlations

In the dynamic world of financial markets, we witness a captivating interplay where numbers mingle with human behavior, creating a complex symphony of coordinated movements. These bustling environments, though seemingly chaotic, reveal deeply interwoven patterns shaped by forces both apparent and concealed. At first sight, the daily fluctuations in stock prices may seem as unpredictable as a tempestuous ocean. However, beneath this surface turmoil lies a realm where discernible patterns form, subtly guided by principles of coordination that dictate the rhythm of market dynamics. This chapter invites you to delve into the intriguing depths of market correlations, where feedback loops, network effects, power laws, and global economic indicators together weave the intricate fabric connecting financial systems worldwide.

Within these financial mazes, feedback loops can amplify emotions, spawning trends that ripple through networks like waves across water. Network effects further intensify these movements, as collective investor behavior can send stock prices soaring to great heights or plummeting to lows. Market volatility, often misconstrued, adheres to power laws, with rare and impactful events shaping the economic landscape. As we navigate these interconnected phenomena, the alignment of global economic indicators becomes evident, providing insights into how distant markets echo each other's shifts and trends. Through this exploration, we unearth the profound wisdom embedded in these patterns, equipping us with the understanding needed to navigate financial markets with increased confidence and insight.

The Role of Feedback Loops in Financial Markets

Feedback loops are essential elements in financial markets, offering a distinctive perspective on how investor behavior interacts with market dynamics. These loops can intensify trends, creating cycles that push prices higher in bullish phases and lower during bearish ones. A classic example is the momentum effect, where increasing stock prices attract more buyers, further driving demand and causing prices to rise even more. This upward feedback can lead to speculative bubbles, such as those during the dot-com boom, where excitement overshadowed sound valuations. Conversely, in downturns, negative feedback loops can heighten selling pressure, leading to steep declines. Grasping these cycles helps investors anticipate market changes and devise strategies to handle volatility.

The advent of advanced analytics and machine learning has transformed how feedback loops in financial markets are identified and managed. By employing algorithms that process vast amounts of data, patterns that traditional analyses might miss can be uncovered. These technologies enable real-time monitoring and adjustment, providing investors with a significant advantage. For example, quantitative hedge funds often use sophisticated models to exploit feedback-induced market anomalies, executing trades rapidly to seize brief opportunities. By integrating such technologies, market players can not only respond to these loops but also predict their emergence, thus securing a competitive edge in the constantly changing financial landscape.

Although feedback loops can drive extreme market behaviors, they also hold potential for stabilization if managed wisely. Central banks and regulatory bodies are crucial in moderating these loops to prevent systemic risks. By adjusting policies related to interest rates and liquidity, they can reduce excessive market fluctuations. The 2008 financial crisis highlighted the need for such interventions, showing how unchecked feedback loops can lead to disastrous outcomes. In response, regulatory frameworks have evolved to include more sophisticated monitoring tools, aiming to preemptively address destabilizing dynamics. This proactive stance not only protects the financial system but also promotes a more resilient economic environment.

Beyond regulatory measures, both individual investors and institutions can employ strategies to lessen feedback loops' impact on their portfolios. Diversification is fundamental, reducing reliance on any one asset class susceptible to feedback-driven volatility. Additionally, applying behavioral finance principles can help investors identify and counteract cognitive biases that worsen feedback loops, such as herd behavior and overconfidence. By fostering a disciplined investment approach and maintaining a long-term perspective, market participants can navigate the ups and downs of financial cycles more effectively, minimizing the effects of short-term disruptions.

Exploring feedback loops in financial markets also provokes a broader reflection on their implications in other areas. How can the insights from these loops deepen our understanding of complex systems in fields like biology or ecology, where similar dynamics are present? Encouraging interdisciplinary dialogue can lead to innovative solutions for challenges that span multiple fields. By inviting readers to consider these broader applications, we not only enhance their grasp of financial systems but also inspire a comprehensive approach to problem-solving, where insights from one domain illuminate pathways in others.

Network Effects and Their Impact on Stock Prices

In the complex landscape of financial markets, the influence of network effects is crucial in determining stock prices. These effects emerge when the value of an asset is impacted by the number and interconnections of market participants. Unlike classic supply and demand, network effects are driven by the principle that increased connectivity among participants enhances asset valuation. For example, social media can dramatically influence stock prices; a single tweet might cascade through investor networks, triggering a collective buying spree. This behavior illustrates the power of collective sentiment and the dynamic nature of interconnected systems.

Recent research emphasizes the growing importance of network-driven dynamics in financial markets. The proliferation of algorithmic and high-frequency trading platforms highlights this trend, as they leverage network effects to execute rapid trades, often amplifying market fluctuations. These systems can identify subtle patterns and correlations, using them to predict and exploit price changes. Their interconnectedness, coupled with the sheer volume of executed trades, can provoke significant price shifts in mere seconds. Consequently, traditional investors must adapt, incorporating real-time data analytics and machine learning tools to keep pace with this swiftly changing environment.

Network effects extend beyond algorithmic trading, affecting the broader market ecosystem and influencing investor behavior. Online trading communities, for instance, have democratized market participation, allowing individual investors to collectively impact stock prices. Platforms like Reddit's WallStreetBets show how coordinated retail investor actions can challenge traditional market expectations, leading to surprising price spikes. This phenomenon underscores the democratizing potential of network effects, empowering smaller investors to wield influence previously reserved for institutional players.

To strategically leverage network effects, market participants need a nuanced understanding of the relationship between connectivity and influence. By analyzing social media trends, sentiment, and network structures, investors can identify emerging opportunities. Incorporating behavioral finance principles can also aid in predicting how news and events might spread through networks, offering insights into potential market movements. A network-focused approach enables traders to capitalize on shifts in market sentiment and dynamics.

As markets evolve, the impact of network effects will likely grow, fueled by technological advancements and increased global connectivity. Investors adept at navigating these complexities will be better positioned to recognize and seize opportunities in an increasingly interconnected world. By fostering adaptability and continuous learning, market participants can thrive amidst the unpredictability introduced by network effects. Encouraging curiosity and exploring unconventional strategies can lead to innovative approaches that harness the full potential of network-driven market dynamics.

Power Laws in Market Volatility and Crashes

Financial markets, often seen as unruly and unpredictable, are actually guided by mathematical principles that can reveal patterns in their fluctuations and downturns. One such principle is the concept of power laws, which describes how both large events and minor occurrences are statistically distributed within markets. Unlike the Gaussian distributions that apply to more predictable phenomena, power laws explain the outsized impact of rare, extreme events like market crashes. This understanding challenges traditional risk management models, encouraging investors and policymakers to recognize the significant influence of these 'black swan' events that, although infrequent, can drastically alter economic landscapes.

Central to this concept is the self-similar nature of power laws, where patterns recur at various scales. This fractal-like characteristic allows for a deeper comprehension of market dynamics, as small fluctuations can escalate into major upheavals. By examining historical data, researchers have found that market crashes often follow the same statistical patterns seen in natural disasters and social phenomena. This cross-disciplinary insight highlights the universality of power laws, prompting a reevaluation of how we interpret and respond to market signals. Advanced algorithms and machine learning techniques now harness these insights, providing sophisticated tools to anticipate potential financial disruptions.

The practical application of power laws extends beyond theoretical considerations, offering strategies for market participants. Investors who understand these principles can better navigate market volatility by diversifying portfolios and adopting strategies that account for the likelihood of extreme events. Furthermore, power laws emphasize the importance of resilience in financial systems. By designing mechanisms that can absorb shocks and maintain stability, institutions can mitigate the cascading effects of market disturbances. This proactive approach not only improves individual success but also enhances the overall stability of global economic systems.

Despite their predictive potential, power laws warrant careful examination and debate. Some experts caution that overreliance on these models might lead to complacency, as they do not consider all factors influencing market behavior. Emerging research suggests integrating power law analysis with behavioral economics and sentiment analysis to create a more comprehensive view of market dynamics. This multidimensional approach could uncover new patterns and interactions that traditional models overlook, fostering a nuanced understanding of how human behavior intertwines with mathematical principles to drive market shifts.

To apply these advanced concepts to everyday financial decisions, consider how power laws impact practical strategies. Investors might test their portfolios against hypothetical 'tail events' to assess resilience. Financial educators can incorporate power law awareness into curricula, equipping future analysts with the tools to interpret complex market data. Policymakers could use these insights to develop regulations that anticipate and cushion against systemic shocks. By embracing the fractal nature of financial systems, individuals and institutions can contribute to a more robust and informed economic landscape.

Synchronization of Global Economic Indicators

The complex interplay of global economic indicators resembles the captivating patterns of synchronization found in nature, where seemingly unrelated elements align seamlessly. This phenomenon is driven by the intricate web of financial systems that connect the world, forming a network of interdependencies. As markets become more interlinked, changes in one region can ripple across continents, leading to widespread economic adjustments. For instance, a shift in one nation's commodity prices might impact inflation rates and interest policies elsewhere, highlighting the fluid movement of economic energy within these networks. This interconnectedness offers both opportunities and challenges, requiring a sophisticated approach to navigate the ever-changing economic landscape.

Recent advancements in data analytics and computational models have deepened our understanding of how these global indicators align. Machine learning algorithms now analyze vast datasets, uncovering hidden patterns that traditional methods might miss. Researchers have discovered how central bank policies, geopolitical events, and even climate anomalies can align economic indicators, revealing the profound interconnectedness of our world. These insights challenge long-standing beliefs about market independence and underscore the need for adaptive strategies that embrace this complexity. By leveraging these insights, policymakers and investors can build resilience against unexpected economic shocks and capitalize on synchronization.

This alignment of global economic indicators is not solely a product of technological progress; it also reflects our human instinct to discern patterns and seek predictability. Investors, policymakers, and economists are motivated by a desire to understand market dynamics, leading to theories and models attempting to grasp the essence of economic alignment. The Efficient Market Hypothesis, for instance, suggests that asset prices fully incorporate all available information, implying a form of coordination in market behavior. While debates about the validity of such models continue, they highlight the enduring fascination with unraveling the factors driving market correlations.

To tap into the potential of these aligned patterns, a multi-dimensional approach that goes beyond traditional economic analysis is essential. This involves integrating insights from fields like behavioral economics, network theory, and complexity science. Such an approach offers stakeholders a comprehensive view of the numerous factors influencing market behavior. Imagine using network analysis to map the intricate connections between key economic indicators, uncovering hidden dependencies and vulnerabilities. This not only enhances understanding but also equips decision-makers with tools to anticipate and mitigate systemic risks, fostering a more stable and prosperous global economy.

In this dynamic economic environment, proactive engagement is crucial. Readers are encouraged to adopt a mindset of continuous learning and adaptability, embracing the ever-evolving nature of global markets. Staying informed about emerging trends and using advanced analytical tools can help individuals and organizations strategically position themselves within the aligned global economy. Consider forming collaborative networks with experts from various fields to foster a culture of knowledge sharing and innovation. Through these efforts, we can collectively navigate the complexities of economic alignment, turning challenges into opportunities and paving the way for a brighter, more interconnected future.

In our exploration of the complex interplay between various synchronization patterns, we encounter a vibrant mosaic of connections—ranging from quantum linkage to biological cycles and market trends. These seemingly unrelated phenomena speak a common language of coherence, where the alignment of individual elements forms resilient entities. Quantum linkage illustrates the profound interconnection between particles over great distances, challenging our notions of individuality and separation. In the biological domain, rhythmic cycles govern life's symphony, from the gentle beat of a heart to the grand cycles of ecosystems, highlighting the interdependence crucial for survival. Similarly, market trends reflect this interconnectedness, mirroring the ebb and flow of human actions and their broader impacts on global economies. These patterns collectively underscore the importance of harmonization as a powerful force shaping both nature and society. With this understanding, we gain the ability to leverage the power of coordinated efforts, motivating us to cultivate unity in our pursuits and appreciate the subtle dynamics of collective interactions. As we move forward to examine the next fundamental pattern, we carry the insight that small harmonies can drive significant change. How can we better align our actions with the world's rhythms to foster meaningful transformation?

Chapter Six
Phase Transitions And Tipping Points

Beneath the serene surface of a frozen lake, a hidden transformation awaits the arrival of spring. As the season whispers its arrival, the once-solid ice begins to crack and dissolve, revealing a world teeming with life below. This natural marvel serves as a metaphor for the profound shifts that occur across various systems when they reach a crucial point. Much like the lake's shift from ice to water, countless processes undergo significant changes when they reach their tipping points. These rapid transformations, known as phase transitions, extend beyond chemistry, resonating through ecosystems, societies, and even the universe itself.

Consider the intricate balance within a thriving ecosystem, where each species plays a vital role in maintaining harmony. A slight change—a drop in temperature, the introduction of a new predator—can trigger a cascade of events, leading to the decline of species populations. Similarly, social movements often lie dormant, fueled by a collective desire for change. When enough voices join together, they reach a tipping point, igniting a movement that can reshape societies. These occurrences highlight a universal truth: the moment when incremental actions or changes shift the balance, leading to significant transformations.

In this chapter, we will delve into the enigma of these phase transitions and tipping points, exploring their presence across various fields. By examining the forces that drive these sudden changes, we gain insight into the fundamental principles that govern our world. From the precise alignment of molecules during chemical changes to the ripple effects within social structures, you will uncover how these patterns are interwoven into the fabric of existence. By understanding these shifts, we not only gain a deeper appreciation for the balance of systems but also learn how to leverage these principles to encourage positive changes in the world around us.

Chemical State Changes

For centuries, scientists have pondered the intricate dance of molecules as they transition between states, unveiling the complex choreography of phase changes. At the core of these transformations lies a network of molecular interactions, deeply intertwined and governed by interdisciplinary principles. As solids dissolve into liquids and liquids vaporize into gases, these state changes mirror the profound interconnectedness in both natural and human systems. This delicate dance of particles is more than just a chemical curiosity; it serves as a metaphor for understanding broader patterns of change. Here, the lines blur between the microscopic and macroscopic, inviting exploration into the universal language of transformation.

Our exploration begins with the hidden forces driving these shifts, such as the energy barriers that must be overcome for reactions to occur. The fragile process of nucleation and growth, where order emerges from chaos, parallels the formation of stars and the evolution of ideas. Critical phenomena and scaling laws offer insight into the predictability of these transformations, hinting at underlying order within apparent randomness. As we delve into these subtopics, we reveal the elegant simplicity behind complex systems, empowering us to recognize and leverage similar patterns in the world around us. This journey through the microcosm not only deepens our understanding of chemical processes but also equips us with insights to inspire meaningful change across diverse areas of life.

Molecular Interactions in Phase Transitions

The captivating realm of molecular interactions during phase shifts unveils the intricate ballet of particles that underlies significant transformations in matter. Central to these changes is the complex interaction of forces among atoms and molecules, which dictates how substances transition from one form to another. Phase changes illustrate how structured order can emerge from apparent chaos, a concept echoed in numerous natural phenomena. Take the example of ice melting into water: as the temperature increases, the strong hydrogen bonds in ice weaken, enabling more fluid molecular movement and resulting in a liquid state. This seemingly common occurrence underscores the profound impact of molecular interactions on material properties.

Advancements in molecular dynamics simulations have revealed the detailed routes molecules take during these shifts. By meticulously tracking atomic movements, researchers have discovered hidden processes, such as transient structures that aid state transitions. These findings not only deepen our understanding of phase changes but also lead to innovative applications, like creating materials with specific melting points or developing pharmaceuticals with exact crystallization features. The ability to control these molecular interactions opens new avenues in materials science, providing tools for designing substances with tailored characteristics.

Outside the lab, the principles of phase shifts resonate with broader scientific pursuits, such as studying supercooled liquids and the mysterious nature of glass transitions. The challenge lies in deciphering the complex molecular choreography that triggers sudden state changes. Using advanced experimental techniques, researchers are investigating the conditions that determine whether a substance crystallizes or remains amorphous, a question relevant to manufacturing and biological stability. This exploration not only enriches our understanding of phase changes but also aligns with the broader quest to comprehend how order arises from disorder in various fields.

The study of phase shifts also invites reflection on the critical phenomena occurring at the cusp of change. As a system nears its critical point, fluctuations in molecular arrangements become notable, showing scaling laws that defy standard expectations. This behavior, marked by a divergence in correlation lengths and susceptibility, mirrors the universal patterns observed in other complex systems, such as financial markets or ecosystems on the brink of collapse. By drawing parallels between these diverse fields, we gain a holistic view of how small-scale interactions can lead to large-scale transformations, offering a perspective on nature's interconnected processes.

For those aiming to harness the power of phase changes in practical applications, a nuanced understanding of molecular interactions is crucial. Whether developing new materials with improved thermal properties or optimizing industrial processes for enhanced efficiency, the insights gained from studying phase shifts offer tangible benefits. By embracing the complexity of these interactions and the emergent properties they confer, innovators can devise strategies that capitalize on the inherent potential of matter to transform. This journey into the molecular realm is not just an academic endeavor but a pathway to finding solutions for real-world challenges, highlighting the significant impact of molecular interactions on the larger world.

Energy Barriers and Activation in Chemical Reactions

Exploring the complexities of energy barriers and activation in chemical processes reveals a realm where tiny interactions lead to significant transformations. Central to these changes is the concept of activation energy, the essential threshold that molecules must reach to trigger a reaction. This is not merely an obstruction but an energetic challenge affecting both the speed and possibility of a reaction. When molecules collide with enough force, they overcome this barrier, resulting in a successful change. Activation energy can be envisioned as a mountain separating reactants from products, scaling which requires sufficient kinetic energy. This foundational concept is crucial across fields like biochemistry, where it regulates enzyme activity, and industrial chemistry, where it informs catalyst design to lower these barriers, accelerating reactions and enhancing efficiency.

Recent advancements have significantly expanded our understanding of activation energy, with cutting-edge research highlighting quantum mechanical effects and molecular orientation's role in reaction dynamics. The traditional view, which focused on energy levels, is now enriched by insights into how molecular arrangement during collisions affects reaction likelihood. Quantum tunneling, where particles pass through energy barriers instead of over them, introduces additional complexity, particularly in low-temperature reactions. This phenomenon challenges conventional views and suggests new pathways, especially in astrochemistry, where conditions differ from Earth's norms.

The relationship between energy barriers and reaction kinetics extends beyond theory and holds practical relevance. By understanding these barriers, chemists can manipulate reaction environments to achieve desired results. Adjusting temperature, pressure, or adding catalysts can reshape the energy landscape to favor particular reactions while hindering others. In pharmaceuticals, this knowledge aids in crafting drugs that selectively interact with biological targets, reducing side effects. Green chemistry also leverages this understanding to minimize harmful byproducts by optimizing reaction conditions.

To grasp these ideas more practically, imagine a bustling street where pedestrians symbolize molecules. The energy barrier resembles a busy intersection that only some can cross depending on the traffic signal—the activation energy. Altering the light's timing or adding a pedestrian bridge (catalyst) can change the flow, akin to how modifying reaction conditions affects molecular interactions. This analogy highlights the value of strategic intervention in chemical processes, showcasing human ingenuity in shaping reaction pathways.

Reflecting on the implications of energy barriers and activation in chemical reactions, we should consider both scientific insights and innovation potential. Mastery of these concepts empowers researchers and practitioners to pioneer solutions that transcend traditional limits, addressing challenges from environmental sustainability to medical advances. The journey through the landscape of chemical transformations is one of continuous discovery, where each insight into molecular dynamics brings us closer to uncovering nature's myriad secrets.

Nucleation and Growth in State Changes

Nucleation and growth in phase transformations illustrate a remarkable interaction between microscopic and macroscopic worlds. Central to this process is nucleation, where small clusters of atoms or molecules form stable structures that serve as seeds for larger formations. These initial clusters, once established, can expand and spread within a medium, facilitating the shift of a system from one phase to another. The creation of these nuclei is shaped by several factors such as temperature, pressure, and the presence of impurities or catalysts, all of which can significantly influence the system's energy dynamics. This delicate interplay determines whether a system remains metastable or transitions into a new phase, such as the crystallization of supercooled liquids or vapor condensing into liquid.

As these nucleated clusters grow, they undergo a dynamic process driven by the diffusion of molecules to nucleation sites, followed by their reorganization into a new phase. The rate of diffusion, the availability of molecular building blocks, and the structural compatibility of these molecules are crucial in dictating the pace and pattern of growth. In some scenarios, growth occurs in a layer-by-layer manner, akin to epitaxial growth in thin films, while in others, it follows more intricate paths, like the dendritic growth common in crystallization. This evolution from nucleation to a well-defined phase vividly demonstrates how order emerges from disorder, propelled by matter's inherent drive to minimize energy and enhance stability.

Recent breakthroughs in nanotechnology and materials science have revolutionized our ability to explore and manipulate these nucleation and growth processes. By meticulously controlling the conditions under which nucleation occurs, scientists can customize material properties at the atomic scale. For example, developing new catalysts relies on understanding and directing nucleation pathways to boost reaction efficiencies. Similarly, designing advanced electronic materials with specific optical or mechanical attributes depends on controlled nucleation and growth processes. These advances not only deepen our understanding of phase changes but also pave the way for innovative applications across various sectors.

The study of nucleation and growth is further enriched by the concept of critical phenomena and scaling laws. These principles provide insights into system behaviors near critical points, where minor fluctuations can trigger significant changes. Grasping these scaling laws is vital for predicting material behavior under varying conditions, such as temperature and pressure, and for crafting processes that leverage these critical transitions for technological benefits. By examining the universal aspects of these phenomena, researchers can identify shared characteristics across diverse systems, from superconductors to biological membranes, offering a unified framework for understanding complex transitions.

Imagining practical applications of these insights prompts us to consider broader implications beyond traditional scientific fields. Picture urban planners drawing inspiration from these natural processes to design cities that adapt and grow sustainably, with infrastructure evolving in response to environmental changes. Or envision harnessing these principles to devise strategies that initiate positive social change by identifying and nurturing the initial "seeds" of progressive movements. By applying the lessons from nucleation and growth, individuals and organizations can create environments where small, intentional actions lead to transformative outcomes, mirroring nature's own graceful transitions.

Critical Phenomena and Scaling Laws in Chemistry

In the fascinating world of chemistry, critical phenomena showcase intriguing and complex transformations occurring under specific conditions. These events are marked by sudden shifts in material properties, typically observed near critical points where distinct phases converge. At these critical junctures, even minor disturbances can cause significant changes, highlighting the delicate equilibrium of these transitions. A quintessential example is the shift between liquid and gas phases, where temperature and pressure align at a critical point, resulting in a supercritical fluid state. In this unique state, traditional distinctions between liquid and gas vanish, unveiling a realm where density fluctuations and molecular interactions undergo remarkable alterations. This intricate interplay of forces reveals the complex nature of phase transitions and provides insights into the subtleties of matter during its most transformative moments.

The scaling laws associated with critical phenomena offer profound insights into how materials behave near these transitional thresholds. These laws describe the changes in various physical properties, such as heat capacity or magnetization, as a system nears its critical point. The concept of universality becomes evident here, showing that diverse systems can exhibit similar scaling behaviors despite differing in microscopic details. This universality enables scientists to predict the behavior of complex materials by drawing knowledge from simpler systems. For example, the principles governing the critical behavior of liquid-gas transitions can be applied to magnetic systems, illustrating the interconnectedness of seemingly disparate domains. This shared mathematical framework not only enhances predictive capabilities but also fosters innovation across disciplines, as researchers draw parallels between fields to uncover hidden patterns.

Recent advancements in computational chemistry and high-resolution imaging have elevated the study of critical phenomena. Researchers now utilize sophisticated simulations to model molecular interactions with exceptional accuracy, while experimental techniques such as neutron scattering and X-ray diffraction provide detailed views of matter undergoing phase changes. These cutting-edge methodologies allow for the exploration of nanoscale processes that underpin macroscopic changes, revealing the complex choreography of atoms and molecules during critical transformations. By understanding these mechanisms, scientists can design materials with tailored properties, paving the way for innovations in areas such as drug delivery, energy storage, and nanotechnology. The ability to manipulate and harness these transitions opens new avenues for technological advancement, highlighting the practical significance of understanding critical phenomena.

As we explore the mysteries of critical phenomena, thought-provoking questions arise, challenging our understanding of the natural world. What governs the emergence of universal behaviors across distinct systems, and how can we exploit these principles to drive innovation? These inquiries encourage reflection on the interconnectedness of scientific disciplines and promote a holistic approach to problem-solving. By embracing diverse perspectives and fostering interdisciplinary collaboration, researchers can uncover new insights into the mechanisms driving critical transformations. This integrated approach not only enhances our comprehension of the fundamental laws governing matter but also inspires novel strategies for addressing some of society's most pressing challenges.

To translate these insights into actionable outcomes, readers are encouraged to apply the principles of critical phenomena to real-world scenarios. Whether optimizing industrial processes, designing resilient materials, or developing sustainable technologies, understanding the intricacies of phase transitions can lead to transformative solutions. By recognizing the potential of small-scale changes to precipitate large-scale impacts, individuals can harness the power of critical phenomena to effect meaningful change. This proactive engagement with the principles of chemistry empowers readers to contribute to the ongoing quest for innovation and discovery, reinforcing the idea that even the most subtle shifts can catalyze profound transformations in our world.

Species Population Collapse

The collapse of species populations reverberates through the fabric of life, exposing the delicate equilibrium that underpins ecosystems. As we delve into this subject, the intricate connections among all living beings become evident, illustrating how the loss of a single species can unleash a chain reaction affecting entire ecosystems. This ripple effect highlights the urgency of understanding and predicting these declines before they happen, emphasizing the necessity for vigilance in safeguarding biodiversity. It's a complex interplay where each species contributes its part, and when one stumbles, the whole system can be thrown off balance.

The early indicators of these profound changes often go unnoticed, yet they hold the potential to prevent ecological catastrophes. Within this elaborate network, certain species serve as vital anchors, sustaining the stability of their environments. These keystone species, like unseen engineers, uphold balance, and their reduction can lead to ruin for other species. To grasp and anticipate these dynamics, mathematical models serve as essential tools, shedding light on population trends and the likelihood of collapse. Human actions, from habitat destruction to climate change, heighten these dangers, but also present opportunities for proactive intervention. By understanding these dynamics, we can devise strategies to protect and restore the fragile balance of nature, ensuring a thriving future for biodiversity.

Identifying Early Warning Signals in Ecosystems

Ecosystems are complex networks of interdependence, with their health often indicated by subtle changes. The concept of early warning signals in these environments is gaining importance as scientists aim to predict major shifts before they happen. One notable signal is critical slowing down, which describes the extended time an ecosystem takes to return to balance after a disturbance. This can be likened to a pendulum that swings back to its resting position more slowly as it nears a tipping point. By tracking variations in recovery times, researchers can identify signs of potential collapse, offering a chance to intervene before irreversible harm occurs.

To detect these signals, researchers increasingly rely on advanced computational models and data analysis techniques. The use of machine learning in ecosystem studies has transformed our capacity to process large datasets, uncovering hidden patterns and relationships. These models can identify small changes in species diversity, migration trends, and resource availability, providing a framework to predict ecosystem health. By leveraging these advanced tools, conservationists gain a deeper understanding of ecosystem functions and potential failures, allowing for proactive measures to protect biodiversity.

Specific examples highlight the effectiveness of these early warning systems. In coral reef ecosystems, slight changes in water temperature and acidity can predict bleaching events, which are devastating for marine life. Monitoring systems that track these variables in real-time have enabled scientists to forecast and mitigate bleaching impacts, safeguarding crucial marine habitats. Similarly, in terrestrial systems, the decline of pollinator populations serves as an early warning of broader agricultural and ecological repercussions. Addressing the factors behind these declines, such as pesticide use and habitat destruction, can prevent cascading effects on food production and biodiversity.

The role of citizen science in identifying early warning signals is invaluable. As technology becomes more accessible, individuals equipped with smartphones and other devices contribute significant data, enhancing the scope of ecological monitoring. This democratization of data collection empowers communities to engage in conservation efforts, fostering stewardship and accountability for local ecosystems. By harnessing the power of collective observation, we can deepen our understanding of environmental changes and their consequences, creating a robust network of informed and active citizens.

Incorporating early warning systems into conservation strategies requires a shift from reactive to proactive approaches. Decision-makers must adopt adaptive management practices that integrate real-time data to guide policies and actions. This involves not only investing in technology but also fostering collaborations among scientists, policymakers, and local communities. By embedding early warning systems into environmental governance, we can improve our ability to anticipate and mitigate ecological crises, ensuring the resilience and sustainability of our planet's diverse ecosystems.

The Role of Keystone Species in Ecosystem Stability

Keystone species are integral to ecosystems, acting as the backbone that maintains ecological equilibrium. These often-overlooked organisms exert a significant influence on their surroundings. For example, the sea otter plays an essential role in preserving the health of kelp forests by controlling sea urchin populations. Without otters, urchins would overwhelm the ecosystem, destroying kelp forests and the diverse marine life they support. This complex network of interactions showcases the critical impact keystone species have on their environments, emphasizing their importance in sustaining biodiversity.

Recent research sheds light on how keystone species exert their influence. In the African savanna, elephants have been found to play a crucial role in shaping plant communities and influencing fire patterns, thus maintaining diverse habitats that support various wildlife. By uprooting trees and trampling vegetation, elephants create open spaces that encourage grass growth, vital for herbivores. This, in turn, supports predators and scavengers, creating a dynamic ecosystem that relies on the presence of elephants. These findings highlight the need to protect these vital species to prevent ecological disruptions.

The disappearance of a keystone species can lead to significant, sometimes irreversible changes in ecosystem dynamics. For example, the decline of jaguars in Central and South American forests has resulted in increased herbivore and smaller predator populations, affecting vegetation and the distribution of other species. These shifts can alter nutrient cycling and soil composition, illustrating how keystone species support ecosystem resilience. Understanding these interactions encourages conservation strategies to preserve these essential organisms and maintain ecological balance.

Advancements in ecological modeling and data analysis have deepened our understanding of the roles keystone species play. Modern research uses network analysis and machine learning to map complex ecosystem relationships, identifying key players that sustain stability. These tools not only enhance our comprehension of current ecosystems but also help predict the effects of losing keystone species. By simulating different scenarios, researchers can devise targeted conservation efforts to prevent ecological collapse, ensuring biodiversity and ecosystem services continue.

Given these insights, practical actions are necessary to strengthen ecosystems by preserving keystone species. Conservationists and policymakers can focus on habitat protection, initiate rewilding projects, and invest in community-based conservation programs, involving local communities in safeguarding these crucial organisms. Encouraging collaboration among ecologists, economists, and social scientists can lead to innovative conservation approaches, ensuring keystone species thrive and continue to weave the intricate tapestries of life. As we face biodiversity challenges, the urgency to recognize and protect keystone species has never been more apparent.

Mathematical Modeling of Population Dynamics

The intricate process of modeling population dynamics in mathematical terms involves a complex interplay of variables that reflect the rise and fall of species within ecosystems. Central to these models is the use of differential equations that map out how populations evolve over time, taking into account factors such as birth and death rates, migration, and environmental influences. The Lotka-Volterra equations offer a foundational perspective on predator-prey interactions, showing how fluctuations in one species can lead to changes in another. Recent advancements have expanded these models, incorporating randomness and spatial variations to better represent the unpredictable and diverse nature of real-world ecosystems.

Recent studies have leveraged computational simulations and machine learning to refine the accuracy and flexibility of population models. By integrating data from satellite imagery and sensor networks, scientists can enhance these models with up-to-date environmental data, capturing the subtle effects of climate change and habitat fragmentation. These innovations help identify crucial tipping points, where minor changes in conditions might cause a cascade of effects, potentially leading to population collapse or recovery. This knowledge is invaluable for conservationists and policymakers aiming to preserve ecological balance and prevent biodiversity losses.

Keystone species play a significant role in these mathematical models, highlighting their outsized impact on ecosystem stability. By analyzing the intricate interdependencies between keystone species and their surroundings, researchers can predict the consequences of their removal or decline, which may ripple through food webs and cause widespread disruptions. For example, a reduction in beaver populations, often considered ecosystem engineers, can drastically change water dynamics and plant communities, illustrating the profound interconnectedness these models capture. Simulating scenarios involving keystone species loss allows stakeholders to develop strategies to strengthen ecosystem resilience.

Beyond theoretical insights, mathematical modeling is a practical tool for devising strategies to combat biodiversity loss. By simulating different intervention strategies, such as habitat restoration or controlled breeding, these models offer a testing ground for solutions before actual implementation. Decision-makers can thereby assess the long-term effectiveness and possible unintended outcomes of their actions, optimizing conservation efforts. Additionally, these models provide a forum for interdisciplinary collaboration, where ecologists, mathematicians, and computer scientists come together to create innovative approaches to sustainable coexistence with nature.

Exploring population dynamics through mathematical modeling invites contemplation of humanity's broader role within ecosystems. By understanding the mathematical foundations of population changes, society can foster a more symbiotic relationship with the natural world, using this knowledge to create environments where both human and non-human life can flourish. As readers engage with these ideas, they are encouraged to consider their impact on local ecosystems and take tangible steps to promote biodiversity, such as supporting conservation projects or advocating for policies that prioritize ecological health. Through this perspective, the intricate dance of population dynamics becomes a catalyst for meaningful change, inspiring a collective commitment to protecting the planet's diverse life.

Human Impact and Mitigation Strategies for Biodiversity Loss

Human activities, such as urban development and large-scale farming, have dramatically transformed natural environments, leading to a decrease in biodiversity that endangers the stability of ecosystems. This human-induced impact disturbs the natural balance, frequently causing drastic declines in species populations. Key factors like habitat fragmentation, pollution, and climate change hasten these disturbances. To combat these effects, innovative conservation approaches are gaining traction. Rewilding projects, for example, focus on reviving ecosystems by reintroducing native species and allowing natural processes to restore equilibrium. These initiatives highlight the potential for humans to positively influence ecosystems by acknowledging the intricate connections between species and their environments.

Pioneering research underscores the critical role of genetic diversity in safeguarding species against environmental shifts. As climate patterns evolve, species with a broad genetic base are better equipped to adapt, thus ensuring their survival and the continued functioning of ecosystems. Tools such as gene banks and assisted migration are modern conservation strategies aimed at sustaining and enhancing genetic diversity. By safeguarding and relocating species to suitable habitats, conservationists seek to strengthen populations against both present and future climate threats. These efforts reflect a forward-thinking approach to biodiversity management, stressing the importance of foresight and adaptability in conservation planning.

Technology plays an increasingly vital role in monitoring and addressing biodiversity loss. Tools like remote sensing and satellite imagery offer extensive data on changes in habitats, enabling timely interventions and informed decision-making. Machine learning algorithms process these large datasets to forecast potential species declines, providing early warnings and setting the stage for targeted conservation efforts. These technological advances represent a move towards data-driven ecological management, facilitating more precise and effective responses to the multifaceted challenges confronting biodiversity today.

Engaging communities and fostering education are crucial to promoting sustainable human-nature coexistence. Empowering local communities to take part in conservation efforts can yield significant advantages, as these groups often hold invaluable traditional knowledge about local ecosystems. Initiatives led by communities, such as sustainable agriculture and eco-tourism, not only safeguard biodiversity but also offer economic incentives for conservation. By incorporating local insights and practices, conservation efforts become more robust and culturally relevant, enhancing their long-term viability and effectiveness.

Exploring alternative conservation models prompts a reassessment of humanity's relationship with the natural world. Concepts like nature-based solutions emphasize collaborating with natural processes to address societal challenges, such as using wetlands for flood management or forests for carbon capture. These approaches acknowledge the intrinsic worth of biodiversity and aim to harmonize human progress with ecological sustainability. By nurturing a deeper appreciation of the symbiotic relationship between humans and the environment, individuals and communities are encouraged to adopt practices that support biodiversity, ultimately contributing to global ecological resilience.

Social Movement Thresholds

Exploring the essence of social movements reveals how collective human behavior can drive transformational change. These movements often ignite from a shared mission, reaching a pivotal point where scattered voices unite into a powerful chorus that influences societal norms. The elegance of this process mirrors natural events, where seemingly small actions can lead to significant transformations. Every individual's contribution, regardless of size, becomes part of a larger wave capable of reshaping society, much like the butterfly's wings triggering a distant storm. Thus, each action we take plays a role in a broader narrative, sparking ripples that lead to substantial shifts.

To grasp the structure of these movements, we delve into the complex interplay of cascading effects and social dynamics. This journey reveals how mathematical models, particularly those based on network theory, provide insight into predicting and understanding changes in collective behavior. These models clarify how connections amplify influence, transforming individual acts into powerful forces for change. By examining the synergy between theory and practical application, the intersection of network theory and social change emerges as a vital field. It not only offers analytical tools but also empowers communities to use these insights for fostering positive transformation, demonstrating the profound impact a well-coordinated network can have on societal evolution.

The Role of Critical Mass in Social Movements

In the sphere of social movements, the idea of reaching a tipping point plays a crucial role in driving change. This threshold signifies the moment when enough individuals embrace a belief or behavior, instigating a broader societal transformation. It's not just about the number of participants, but also the quality and interconnectedness of these people within social networks. Recent research highlights how linked groups, rather than solitary individuals, act as change agents, magnifying the effects of their actions. This connectivity resembles a network of neurons firing together, with each person contributing to a shared momentum that advances the movement.

Understanding the mechanics of reaching this tipping point reveals that timing and context are vital. The success of a movement often depends on aligning social, political, and cultural conditions that are conducive to new ideas. For example, the civil rights movement in America gained traction not only due to the sheer number of participants but also because of a socio-political environment that was increasingly open to change. The movement reached a turning point when a critical mass of people, united by common grievances and aspirations, came together to confront the status quo.

Advanced research in this area frequently uses complex mathematical models to forecast when a social movement might hit its tipping point. These models borrow from physics and biology, seeing social movements as dynamic systems governed by the same principles as natural phenomena. By applying concepts like percolation theory and bifurcation analysis, researchers can pinpoint the factors that most significantly influence a movement's path. This modeling not only deepens our comprehension of social dynamics but also provides activists and policymakers with tools to strategically cultivate emerging movements.

From another angle, the influence of digital platforms in achieving a tipping point is significant. As online spaces become more central to social interaction, they offer fresh opportunities for movements to gain momentum. Social media platforms, with their extensive networks and rapid information-sharing capabilities, act as accelerators, allowing movements to reach a tipping point faster than before. This democratization of information amplifies diverse voices, potentially shortening the time needed to achieve a turning point. However, this also adds complexity, as these platforms can also be used to spread misinformation or hinder progress.

While we often associate the idea of a tipping point with large-scale movements, its principles are applicable to smaller initiatives too. Individuals and organizations aiming for change can focus on building strong, interconnected networks and fostering conditions favorable to growth. By grasping the dynamics of reaching a tipping point, they can plan their efforts strategically, ensuring their actions contribute to a larger wave of change. This approach not only empowers grassroots efforts but also instills a stronger sense of agency among participants, encouraging them to view their contributions as essential to the movement's success.

Analyzing Cascading Effects and Social Dynamics

Exploring the complex interplay of cascading effects in social movements reveals the deep connection between individual actions and collective dynamics. These cascades, much like dominoes falling in succession, occur when a single change sparks a series of reactions, leading to significant societal transformations. Central to this process is the interaction between personal behaviors and societal norms, where momentum builds as more people adopt new behaviors or beliefs. This isn't just a theoretical idea but a tangible force that has historically fueled revolutions, policy reforms, and cultural changes. By examining these cascading effects, we can see how isolated actions contribute to a larger wave of social change, highlighting the crucial role each participant plays in this transformative network.

Recent research underscores the importance of thresholds in these cascading processes. When a certain number of individuals within a network embrace a change, it triggers a chain reaction that alters the entire system. This critical juncture, often termed a tipping point, is vital in social dynamics. It's fascinating how digital platforms have accelerated these processes, as information spreads quickly in our hyper-connected world. Online communities can reach their tipping points faster than traditional societies, illustrating the power of digital networks in shaping contemporary social movements. The speed and reach of these networks highlight the potential of collective action, driven by both physical and virtual interactions.

Mathematical modeling offers a compelling perspective on these dynamics, using complex network theory and computational simulations to predict how changes in individual nodes—people or entities—affect the entire network. This approach not only deepens our understanding of social movement dynamics but also helps strategize for positive societal impact. For example, identifying key influencers within a network can enhance the reach and impact of a movement, as these individuals act as catalysts for wider adoption. This strategic insight allows organizers to focus their efforts more effectively, leveraging the natural flow of information within networks to advance their causes.

The intersection of network theory and social change serves as fertile ground for innovation and discovery. This confluence reveals how structural properties of networks, such as connectivity and resilience, affect the success of social movements. By analyzing these properties, one can uncover patterns that either facilitate or hinder the spread of transformative ideas. This knowledge not only enriches theoretical understanding but also provides practical tools for those aiming to inspire change. By strategically building and nurturing networks, individuals and groups can enhance their capacity to drive meaningful societal shifts, turning abstract concepts into actionable strategies.

Reflect on the potential of cascading effects and social dynamics to reshape the world. Imagine a scenario where small, deliberate actions ripple through society, generating waves of change that cross geographical and cultural boundaries. Consider the power of a single voice, amplified by the resonance of a network, to change the course of history. How might you contribute to such a movement, using the principles of cascading effects to champion causes you care about? By embracing these insights, you can be more than an observer; you can be an architect of change, orchestrating the symphony of social dynamics to create a harmonious and just world.

Mathematical Modeling of Collective Behavior Shifts

Mathematical modeling is a potent tool for deciphering the complexities of how collective behaviors evolve within social movements. Central to these models is their capacity to simulate and forecast how individual actions coalesce into a unified force that can drive societal change. One pivotal concept within these models is "critical mass," which denotes the point at which enough individuals adopt a new behavior or idea, triggering a rapid spread of change. This process resembles a snowball gaining speed and size as it rolls downhill, eventually reaching a tipping point where change becomes inevitable.

A notable model that encapsulates this dynamic is the threshold model of collective behavior. It suggests that individuals have different participation thresholds based on how many others are involved in a movement. By quantifying these thresholds and simulating interactions within a network, researchers can pinpoint conditions under which a movement is likely to thrive. These models demonstrate that heterogeneous networks, characterized by diverse connections and influence patterns, are particularly effective at catalyzing swift behavioral shifts. This highlights the importance of fostering diverse and inclusive networks to bolster the success of social movements.

In recent years, advances in computational social science have led to sophisticated agent-based models, enabling more detailed simulations of social dynamics. These models surpass basic assumptions by integrating complex variables such as individual motivations, social influence, and environmental factors. For instance, an agent-based model might explore how information disseminates via social media, emphasizing the role of influential nodes, or "hubs," in amplifying a movement's reach. By grasping these dynamics, activists can strategically target key influencers to maximize impact and galvanize collective action.

While mathematical models offer valuable insights, they also prompt ethical questions about predicting and influencing collective behavior. Balancing the use of these tools for positive change with respecting individual autonomy is crucial. Engaging with diverse perspectives and ethical considerations ensures that models are both accurate and socially responsible. By promoting open dialogues and interdisciplinary collaboration, scholars and practitioners can refine these models to better serve the public good, empowering rather than manipulating.

To transform these theoretical insights into practical strategies, readers can use modeling techniques to anticipate potential obstacles and opportunities in their advocacy efforts. By identifying the critical mass needed for change and understanding network dynamics, individuals and organizations can design targeted interventions that harness the power of collective behavior shifts. Simulating scenarios can also provide valuable foresight, allowing for adaptive strategies in response to changing social landscapes. In this way, readers gain not only theoretical knowledge but also actionable insights that can drive meaningful progress in their causes.

Intersection of Network Theory and Social Change

Network theory offers a revolutionary perspective for understanding the complexities of social change. At its essence, this approach reveals how individuals within a community form intricate webs of influence, facilitating the rapid spread of ideas and behaviors. These networks are dynamic; they evolve as new links are made and old ones fade, guiding the path of social movements. The balance of nodes and connections can dictate whether a movement gains momentum or dissipates. By analyzing these networks, one can identify trends and anticipate the success of movements striving to move from obscurity to broad recognition.

Central to grasping how movements reach critical mass is the notion of social contagion, which describes the viral spread of behaviors and ideas across a network, similar to the spread of contagious diseases. This concept underscores the significance of key influencers or 'hubs' within a network, whose endorsement of a movement can significantly boost its proliferation. Recent studies emphasize the crucial role these hubs play in expanding the reach and impact of social movements. For example, the Black Lives Matter movement effectively utilized key figures on social media to spark global awareness and participation. By tactically engaging these influential nodes, movements can harness network dynamics to achieve a pivotal tipping point.

Exploring further into how network structure impacts social change, one encounters the intriguing phenomenon of small-world networks. These are characterized by short path lengths and high clustering, enabling quick information exchange and coordination among participants. Such structures are common in societal settings where geographical and social barriers are reduced. The Arab Spring, for instance, illustrated the power of small-world networks in organizing large-scale protests and maintaining momentum across varied regions. Understanding these networks' architecture equips change-makers with strategies to optimize connectivity and resilience against external forces.

The intersection of network theory and social change also paves the way for innovative modeling of shifts in collective behavior. Advanced computational simulations empower researchers and activists to explore theoretical scenarios, examining different strategies for movement propagation. These models integrate variables like network density, node influence, and outside interventions, offering a platform to refine approaches to real-world challenges. By simulating various configurations and outcomes, practitioners can foresee obstacles and refine their tactics to achieve desired social changes. This predictive capability is vital in navigating the intricacies of contemporary social movements.

To effectively utilize network theory in promoting social change, it is essential to adopt a multifaceted viewpoint that considers both the opportunities and challenges. While networks can amplify voices and enable swift mobilization, they also pose problems such as echo chambers and misinformation. Navigating these complexities necessitates a deep understanding of the interplay between technology, human behavior, and societal structures. By fostering an environment that encourages open dialogue and critical thinking, individuals and organizations can harness the transformative power of networks to drive meaningful change. Experimentation and adaptability remain critical, as the landscape of social movements constantly evolves, demanding innovative approaches to address modern issues.

In concluding our journey through phase transitions and tipping points, the intricate links between various phenomena stand out vividly. Whether observing the dramatic changes in chemical states, the abrupt decline of species, or pivotal moments in social movements, these patterns underscore how minor tweaks can catalyze transformative shifts. By grasping these dynamics, we equip ourselves to foresee and shape crucial turning points, transforming looming crises into avenues for growth and innovation. The essence of phase transitions teaches us that even the most stable systems can be susceptible to change, emphasizing the need for constant awareness and adaptability. This chapter challenges us to appreciate the fragile equilibrium within complex systems and urges us to act with intention in our daily activities, recognizing our potential to drive significant global transformations. As we move forward to the next chapter, our quest to discover more connections between the micro and macro continues, enriching our understanding and amplifying our capacity to foster meaningful change.

Chapter Seven

Wave Propagation Dynamics

Deep within the Amazon rainforest, the subtle flutter of a butterfly's wings stirs the air. This seemingly trivial motion initiates a sequence of events that will resonate far and wide, reshaping destinies across the planet. This is not mere poetic fancy; it serves as a vivid demonstration of how signals and energy traverse through our universe—a captivating journey through which information and influence spread, often leading to remarkable outcomes. As we delve into this chapter, envision the waves of transformation that arise from a single act, weaving through layers of intricacy to yield results that are both expected and surprising.

The study of wave dynamics is essential to comprehending how the universe communicates within itself. From the mysterious interplay of quantum waves that dictate the behavior of particles to the swift dissemination of ideas across digital networks, these patterns are the threads that compose the essence of existence. As we venture into these ideas, we'll reveal the hidden structure within chaos and the steady pulse beneath the surface of randomness. Whether it's the transmission of a transformative idea through societies or the viral spread of a meme, wave transmission offers a perspective to understand the intricate web of connections that link all things.

These rhythmic patterns bridge the gap between the very small and the vast, uncovering the harmonious symmetry governing natural phenomena and human creativity alike. By grasping the fundamentals of wave dynamics, we unlock the ability to shape systems both large and small. As we embark on this journey, reflect on the waves you can generate, the ripples you can set in motion, and the change you can inspire. This chapter encourages you to look beyond the immediate, to appreciate the vast network of interactions that connect us all, and to use this understanding as a catalyst for positive change.

Quantum Wave Functions

Imagine a world where reality is sculpted by waves of potential, each one a mathematical masterpiece crafted in the subatomic realm. In this microscopic arena, quantum wave functions take the lead, guiding the probabilities that determine the behavior of particles. These functions, with their intricate mathematical foundations, reveal a strange yet captivating landscape of quantum mechanics. Here, particles exist in multiple states until observed, and entanglement forms invisible links between distant objects. As we venture into this exploration, we peer through the lens of quantum theory, aiming to unravel the mysteries of superposition and the collapse of wave functions—concepts that challenge our classical grasp of reality.

This journey through the quantum world not only enriches our understanding of the unusual yet fundamental principles that steer the universe but also provides us with sophisticated computational tools to model these phenomena. By delving into the mathematical structures of quantum wave functions, we unlock new possibilities where the interconnected dance of particles could transform computing and communication. In this realm of endless potential, each subtopic unfolds like a chapter in a grand story, illuminating the intricate interplay of particles and waves. Whether it's the surreal implications of superposition, the deep connections forged by entanglement, or the innovative techniques for simulating these quantum behaviors, we stand on the verge of new insights, ready to harness these principles in ways that transcend conventional boundaries.

The Mathematical Foundations of Quantum Wave Functions

Quantum wave functions underpin the essence of quantum mechanics, providing a mathematical framework for understanding the probabilities linked to a particle's position and momentum. Represented by the Greek letter psi (Ψ), these functions illustrate how particles remain in a potential state until measured. Far from being abstract, the wave function is essential for capturing the probabilistic nature of quantum events, indicating where a particle might be found. This challenges classical ideas, inviting exploration into the complex layers of quantum reality where probability replaces certainty, and observation influences outcomes.

In quantum mechanics, superposition highlights the remarkable behavior of particles existing in multiple states simultaneously. This principle is fundamental to technologies like quantum computing, where qubits use superposition to execute complex calculations rapidly. The mathematical depiction of wave functions captures superposition by assigning coefficients to describe the amplitude of each state. As research progresses, it uncovers the intricate balance between superposition and coherence—maintaining quantum states without interference—as crucial for advancing quantum technologies. Beyond computation, this has potential implications for fields like cryptography and materials science, where controlling quantum states could lead to breakthroughs in encryption and novel material development.

Entanglement, a core concept of quantum mechanics, creates connections between particles that defy classical limits. When particles are entangled, a change in one affects the other instantaneously, regardless of distance. This phenomenon, often called "spooky action at a distance," is represented mathematically within the wave function framework. The wave function collapses upon measurement, resolving entangled states into definite results. Current research is expanding the potential of entanglement, especially in quantum communication and teleportation, where instantaneous information transfer could redefine connectivity and security. Understanding wave function collapse sheds light on the shift from potential to actual, offering insight into the universe's fundamental workings.

Advanced computational methods have become invaluable in modeling quantum wave functions, allowing scientists to simulate intricate systems with great accuracy. Techniques such as density functional theory and quantum Monte Carlo simulations enable exploration of systems that traditional methods cannot handle. These computational advances are essential in fields like quantum chemistry and condensed matter physics, facilitating predictions of molecular behavior and the study of exotic states of matter. By incorporating these cutting-edge methods, researchers are revealing the complexities of wave functions with unprecedented clarity, paving the way for practical applications and new technologies grounded in quantum principles.

The study of quantum wave functions also prompts reflection on the philosophical aspects of quantum mechanics—how its probabilistic nature challenges traditional ideas of determinism and causality. This exploration illuminates the intricate fabric of quantum mechanics, where elements like superposition and entanglement weave a narrative of possibility and connection. By contemplating these concepts, readers are encouraged to rethink the very essence of reality and consider the broader implications of quantum mechanics on our understanding of the universe. Engaging with these ideas not only deepens one's grasp of quantum phenomena but also inspires a reevaluation of the role of observation and knowledge in shaping the world around us.

Analyzing Superposition and its Impact on Quantum States

Quantum superposition is a captivating yet perplexing concept within quantum mechanics. It allows a quantum system to exist in multiple states simultaneously until an observation forces it into one state. This principle not only challenges traditional views but also paves the way for groundbreaking technologies such as quantum computing. In these systems, qubits exploit superposition to perform calculations at speeds unattainable by classical bits, holding promise for revolutionizing fields like cryptography and complex simulations. As researchers delve deeper into superposition, they discover subtleties that may redefine the boundaries of computation and information theory.

A significant outcome of superposition is its contribution to quantum parallelism, enabling a quantum computer to execute numerous calculations concurrently. This stems from the distinctive manner in which quantum states can represent information, allowing for the simultaneous pursuit of multiple solutions to a problem. This capability is not merely theoretical; advancements in quantum processors have demonstrated impressive speedups in specific tasks. Shor's algorithm, for example, highlights how superposition can efficiently factorize large numbers—a task that stumps classical computers. This scenario illustrates superposition's potential to boost computational efficiency and challenge current encryption methods, urging a reevaluation of digital security standards.

Superposition is also intricately linked with interference, where probability amplitudes from different states combine to affect observable outcomes, either constructively or destructively. This phenomenon is crucial in quantum experiments, such as the renowned double-slit experiment, which vividly showcases wave-particle duality and the probabilistic essence of quantum mechanics. By mastering interference patterns, scientists design experiments that probe the very essence of reality, questioning the nature of observation and measurement. These insights hold profound significance for theoretical physics and the development of experimental techniques that push precision and control boundaries.

Beyond the laboratory, exploring superposition's implications extends into the philosophical domain, challenging established ideas about reality and observation. The many-worlds interpretation suggests that all possible outcomes of quantum measurements are realized in parallel universes, offering a radical view on existence. While debates over interpretations persist, they inspire new inquiries and foster a deeper understanding of the quantum landscape. These discussions emphasize the interplay of science and philosophy, urging contemplation of the broader implications of quantum principles on our understanding of the universe.

For those eager to leverage superposition, practical steps include staying informed about the latest advancements in quantum technology. Engaging in interdisciplinary collaborations can spark innovative solutions by merging insights from physics, computer science, and engineering. As quantum technologies inch closer to practical applications, professionals can prepare by exploring how quantum principles might reshape their fields, from optimizing logistics with quantum algorithms to refining machine learning models. A proactive approach to grasping and applying superposition can empower individuals and organizations to navigate and contribute to the evolving quantum revolution, ultimately shaping a future brimming with possibilities.

Quantum Entanglement and Wave Function Collapse

Quantum entanglement, a fundamental aspect of quantum physics, defies traditional ideas of separateness. In this extraordinary occurrence, two or more particles become linked so that the condition of one immediately affects the other, regardless of how far apart they are. This contradicts classical ideas of locality and points to a complex web of connections at the subatomic level. The effects of this entanglement are significant, impacting areas like cryptography and quantum computing. For example, it underlies quantum key distribution, which promises unparalleled security in data transmission. By integrating this concept into practical uses, we envision a future where communication is not only swifter but also more secure.

The collapse of a quantum wave function, often initiated by observation, signifies a shift from possibility to reality. Prior to this collapse, a particle exists in a superposition of states, represented by a wave of probabilities. Measurement forces this wave to select a definite state, a process that remains enigmatic and contentious. This pivotal transition in quantum mechanics sheds light on the influence of observation on reality. Advanced research delves into decoherence, where environmental interactions lead to wave function collapse, indicating that the line between quantum and classical realms is more fluid than once thought. These studies not only increase our comprehension of quantum mechanics but also pave the way for technological breakthroughs utilizing these principles.

Progress in quantum entanglement is transforming computational methods. Quantum computers, which exploit entangled states, hold the promise of resolving problems that classic computers find unsolvable. Algorithms like Shor's, used for factoring large numbers, illustrate exponential improvements over standard techniques. This surge in computational power could revolutionize industries, from pharmaceuticals to finance, enabling complex calculations to be executed with unmatched efficiency. The incorporation of entangled quantum systems into computing frameworks might redefine problem-solving approaches, urging us to reevaluate digital transformation possibilities.

Consider the philosophical ramifications: if entanglement and wave function collapse alter our perception of reality, how might these ideas affect our view of interconnectedness in daily life? The resemblance between quantum interconnections and social networks becomes clear when we see how actions echo across different areas. Just as entangled particles show a hidden unity, human actions influence social structures, affecting behaviors and outcomes in unexpected ways. This analogy prompts us to reflect on our role in larger systems, encouraging us to consider how small, deliberate actions can lead to significant change.

Exploring the intricacies of quantum entanglement and wave function collapse uncovers how these phenomena not only challenge scientific paradigms but also provide practical insights. By embracing the non-intuitive nature of quantum physics, we unlock new strategies for problem-solving and innovation. The most profound takeaway may be the reminder of our connectedness, both at the quantum level and within society. Readers are inspired to adopt a quantum perspective in their pursuits, fostering curiosity and openness to new possibilities that go beyond conventional limits.

Advanced Computational Techniques for Modeling Wave Functions

In the complex field of quantum mechanics, cutting-edge computational techniques for simulating wave functions offer intriguing insights into the behavior of subatomic particles. These methods have advanced considerably, with quantum algorithms and simulations leading current research efforts. For example, quantum computing has introduced innovative ways to simulate wave functions, using qubits to process vast data sets simultaneously. This functionality allows scientists to model intricate quantum systems with unparalleled precision, shedding light on phenomena like quantum tunneling and entanglement. Merging machine learning with quantum simulations further refines these models, enabling algorithms to adapt based on new data and unlocking novel pathways for discovery.

Recent progress in the area includes the application of density functional theory (DFT) for precise modeling of quantum wave functions. DFT provides a framework for calculating the electronic structure of complex systems, aligning theoretical predictions with experimental data. Utilizing these techniques, researchers gain deeper insights into atomic-level chemical reactions, predict properties of emerging materials, and support fields like nanotechnology and materials science. This blend of computational modeling and empirical data not only advances theoretical understanding but also fuels technological innovation, deepening our grasp of the quantum realm.

Simultaneously, the rise of hybrid quantum-classical algorithms presents a promising strategy for tackling complex quantum systems. These algorithms exploit the strengths of both quantum and classical computing, optimizing resources and improving efficiency. By assigning specific tasks, such as simulating quantum circuits, to quantum processors, while using classical computers for error correction and post-processing, researchers achieve a harmonious balance that enhances wave function modeling. This collaborative approach opens doors to solving previously unsolvable problems, expanding the limits of what can be computed and understood in quantum physics.

Exploring varied perspectives within this domain, some researchers advocate for holistic modeling approaches that incorporate environmental factors into simulations. This perspective challenges conventional models that often isolate systems from their surroundings. By examining the interaction between quantum systems and their environments, scientists can develop more comprehensive models that reflect the complexity of real-world scenarios. This approach not only enriches our understanding of quantum phenomena but also underscores the interconnectedness of quantum systems with the broader universe, offering a wider canvas for exploration.

Thought-provoking scenarios arise when considering the practical applications of these advanced computational techniques. Imagine a future where quantum simulations are pivotal in drug design by accurately modeling molecular interactions at the quantum level, leading to more effective treatments with fewer side effects. Alternatively, consider the potential for quantum models to revolutionize renewable energy by predicting the behavior of new photovoltaic materials, enhancing their efficiency and sustainability. By staying at the forefront of computational advancements, researchers and practitioners can harness these techniques to drive innovation, foster progress, and contribute to a deeper understanding of the universe.

Information Spreading Patterns

Imagine a world where information flows as effortlessly and rapidly as light, reaching every part of our interconnected globe. In this dynamic environment, the way information spreads is as vital as the information itself. Each data fragment—whether a trending meme, a health alert, or a cultural shift—moves through networks, leaving a lasting imprint on society. The spread of this information is not random; it follows detailed patterns akin to the graceful movement of waves in nature. By deciphering these patterns, we unlock the potential to use them for innovation, connection, and transformation.

Within this complex web of information exchange, social networks play a pivotal role, accelerating the dissemination of ideas and trends at remarkable speeds. These digital pathways mirror the intricacies of natural systems, where seemingly chaotic interactions lead to emergent communication patterns. As we explore the mathematical models that predict the flow of information and strategies to control or enhance its reach, an intriguing landscape unfolds. Here, cultural norms are not just shared but evolve, influenced by the same forces that shape natural phenomena. Viewing this process through the lens of wave dynamics provides insights into how information saturates our world, offering routes to navigate and influence the tides of change.

Viral Memes and the Mechanics of Cultural Transmission

Viral memes offer a captivating insight into the complex mechanisms of societal transmission. These memes, transcending mere online phenomena, encapsulate the dynamics of how ideas, behaviors, and styles spread across communities. Central to this is replication, where a meme must engage and be shared by individuals, akin to genetic traits passed through generations. Studies in memetics—an interdisciplinary field linking anthropology, sociology, and cognitive science—reveal how memes use emotional appeal and cognitive biases to boost their spread. Grasping these principles allows one to see how cultural artifacts embed in collective consciousness, influencing societal norms and values.

Mathematical modeling of meme transmission provides profound insights into the dynamics of information spread. Using models similar to those in epidemiology, researchers simulate idea proliferation within populations, paralleling pathogen spread. These models consider factors like network topology, transmission rates, and individual susceptibility to novel ideas. Computational social science highlights how network structures affect the speed and reach of meme dissemination. By evaluating these parameters, strategies can be developed to either enhance beneficial memes or curb harmful ones. This blend of mathematics and social science paves the way for predictive analytics in cultural evolution, enabling a proactive approach to shaping public discourse.

Social networks play a crucial role in meme propagation, accelerating information diffusion. Platforms such as Twitter, Facebook, and TikTok demonstrate how digital ecosystems facilitate rapid idea exchange across vast geographical and cultural divides. The architecture of these networks, characterized by clusters and nodes, shapes meme trajectories. Research in network theory underscores the influence of key individuals—those with significant reach and credibility—in speeding up meme adoption. By engaging these pivotal figures strategically, movements and campaigns can achieve viral status, effectively sparking social change. Understanding social network dynamics empowers digital marketers and equips activists and educators with tools to raise awareness and engagement on important issues.

Complex systems theory provides an enriched view of communication patterns, emphasizing the non-linear interactions driving meme transmission. Unlike linear systems, where outputs directly correspond to inputs, complex systems exhibit emergent behavior, where collective actions lead to new phenomena. In cultural transmission, this means spontaneous formation of meme trends and virality. Current research explores how feedback loops within social networks create self-organizing patterns, where memes evolve and adapt based on audience interaction. Understanding these dynamics offers strategic advantages in crafting messages that deeply resonate, fostering a sense of community and shared identity among diverse groups.

Exploring viral memes and societal transmission imparts valuable lessons for those aiming to harness the power of ideas for positive change. Recognizing the patterns and principles that govern meme dynamics allows crafting messages that captivate attention and inspire action. This involves appreciating the audience's cultural context, emotional triggers, and cognitive tendencies. Practical strategies, like storytelling techniques or visual symbolism, amplify meme potency, ensuring they thrive in the cultural landscape. By embracing the science of memetics, one can contribute to a tapestry of ideas enriching human experience and driving meaningful change globally.

Mathematical Models of Epidemic Spread and Containment Strategies

The complex interplay between infectious diseases and their containment strategies is a fascinating area where mathematical models shed light on the dynamics of epidemic spread. These models, often based on differential equations, help unravel the intricacies of pathogen transmission within communities. The Susceptible-Infectious-Recovered (SIR) model is a foundational tool that breaks down an epidemic's progression into distinct phases. This model, along with its variations, allows researchers to forecast outbreak paths, pinpoint critical herd immunity thresholds, and assess the effectiveness of various intervention measures. By simulating multiple scenarios, scientists can evaluate the success of vaccination efforts, quarantine protocols, and public awareness campaigns, providing strategic tools to combat the relentless spread of infectious agents.

Recent advancements in computational technology and data analytics have propelled epidemic modeling into a new era of accuracy. Machine learning algorithms now augment traditional models by incorporating real-time data from diverse sources like social media, travel trends, and weather patterns, offering more sophisticated predictions. These advanced models can account for the random nature of disease spread, including factors like superspreading incidents and asymptomatic carriers. For instance, during the COVID-19 pandemic, machine learning improved the ability to predict outbreak hotspots and tailor containment measures to specific areas, showcasing the adaptability of these models to ever-changing situations.

While models provide valuable insights, they also highlight the complexities of implementing effective containment strategies. The relationship between human behavior and policy adherence adds layers of unpredictability. The success of interventions often relies on public cooperation, making behavioral sciences a crucial element of epidemic response. Models that integrate psychological and social factors offer a more comprehensive view, acknowledging that the success of containment efforts is closely tied to the public's perception and acceptance of these measures. This interdisciplinary approach fosters a deeper understanding of how societal dynamics impact an epidemic's course, empowering policymakers to design strategies that resonate with various communities.

Exploring the innovative realm of network-based models reveals how social connections can either facilitate or impede the spread of infectious diseases. These models map the complex web of human interactions, identifying key individuals or 'super-spreaders' whose behavior significantly impacts transmission dynamics. By targeting these individuals or optimizing social network structures, interventions can be crafted to effectively break transmission chains. This approach highlights the importance of detailed data and a nuanced understanding of social dynamics in creating interventions that are both effective and minimally disruptive.

To translate mathematical insights into actionable strategies, interdisciplinary collaboration is essential. Epidemiologists, data scientists, behavioral experts, and policymakers must work together to bridge the gap between theoretical models and practical applications. By fostering a collaborative environment, these diverse perspectives can merge into innovative containment strategies that are adaptable and robust. As the field continues to evolve, the integration of cutting-edge research, advanced computational tools, and a nuanced grasp of human behavior will remain crucial in crafting effective responses to future health challenges. Through this collaborative and dynamic approach, mathematical models of epidemic spread not only clarify the path of contagion but also chart a course toward resilient and proactive public health strategies.

The Role of Social Networks in Accelerating Information Diffusion

Social networks, whether digital or face-to-face, are potent channels for rapid information exchange. Functioning like complex webs, each node symbolizes a person or organization, with connections forming paths for information flow. These networks can quickly amplify messages, as a single piece of information can travel through numerous connections, reaching large audiences swiftly. This is especially evident online, where platforms like Twitter, Facebook, and Instagram enable the viral spread of news, ideas, and cultural memes. Research shows that the structure and density of these networks play a significant role in the speed and reach of information spread. Networks with high clustering and short paths tend to expedite the diffusion process, highlighting the importance of network design in the dynamics of information flow.

Recent studies in network theory emphasize the role of certain individuals, known as "influencers" or "super-spreaders," in speeding up information diffusion. These individuals have a large number of connections, giving them extensive reach and influence. Their impact is not just due to connectivity but also their credibility and the relevance of their shared information. Targeting these key nodes strategically can greatly enhance the effectiveness of information campaigns, whether for public health messages, social movements, or marketing efforts. Understanding the traits of these influential nodes and making use of their potential can boost dissemination strategies.

The spread of information within social networks is also shaped by the balance between homophily and heterophily. Homophily, where individuals connect with similar others, can create echo chambers that reinforce information within a closed group. In contrast, heterophily, the attraction to diverse connections, can introduce new ideas and spread information across various communities. Balancing these forces is essential for broad information dissemination. Encouraging diverse connections can break down silos and promote the exchange of ideas, leading to stronger and more resilient information networks.

In exploring advanced insights, the role of algorithms in social networks is crucial. Algorithms determine content visibility and reach, often favoring engagement over informational value. This can lead to the spread of sensational or polarizing content, overshadowing nuanced or factual information. Researchers are working on designing algorithms that promote trustworthy and diverse information, improving the information landscape's quality. By prioritizing content that connects different network clusters, algorithms can foster dialogue and understanding, counteracting fragmentation in digital spaces.

Considering the future of information diffusion in social networks, it's vital to ponder the ethical implications of these dynamics. The power of networks to influence public opinion and behavior highlights the need for responsible management and regulation. Important questions arise: How can we ensure that information systems support democratic discourse and collective well-being? What safeguards can prevent the manipulation or misuse of these powerful channels? By addressing these challenges and responsibly harnessing social networks, we can create environments where information flows freely, empowering individuals and communities for positive change.

Complex Systems and the Emergence of Communication Patterns

Within the intricate dynamics of complex systems, communication frameworks emerge with a sophistication that conceals their underlying intricacies. These frameworks dictate the movement of information across extensive networks, uncovering the concealed structures that both facilitate and restrict message transmission among interconnected entities. Fundamentally, these systems are characterized not only by their individual elements but also by the interactions and connections formed between them. Recent research has demonstrated the appearance of such emergent communication frameworks in various scenarios, from the synchronized chirping of crickets to the swift dissemination of technological advancements across global markets. Each instance provides insight into the dynamic processes shaping information transmission and transformation within a network.

Grasping the mechanisms behind these frameworks necessitates a multidisciplinary approach, integrating insights from network theory, cognitive science, and information technology. Take, for example, the communication in ant colonies via pheromone trails. This seemingly straightforward communication method actually represents a complex system balancing exploration and exploitation, enabling the colony to efficiently allocate resources and react to environmental changes. Such cases highlight principles of self-organization and adaptation, where local interactions lead to global behaviors that are often more efficient and resilient than those in human-designed systems.

Feedback plays a crucial role in these systems. Positive feedback can amplify a message, resulting in viral content spread on social media platforms, whereas negative feedback can stabilize communication, preventing system overload and ensuring sustainability. Contemporary research in artificial intelligence and machine learning is beginning to model these feedback mechanisms to bolster the robustness of communication networks, drawing inspiration from biological systems that have evolved over vast timeframes. By leveraging these insights, developers aim to create algorithms that replicate the efficiency and adaptability found in nature, paving the way for innovations in digital communication technologies.

Emerging technologies are also reshaping our understanding of communication frameworks. Blockchain, for instance, is transforming information verification and sharing across decentralized networks, providing a new perspective on trust and consensus in complex systems. Through the lens of complex systems theory, researchers are discovering new insights into how distributed networks can maintain integrity and security without central oversight. This has significant implications for everything from financial transactions to the distribution of digital content, highlighting the potential of complex systems to revolutionize traditional communication models.

As we contemplate the future of communication within complex systems, several intriguing questions arise: How can we harness emergent principles to design more efficient communication networks? What lessons can we draw from natural systems to enhance human-engineered communication platforms? By exploring these questions, we not only deepen our understanding of complex systems but also equip ourselves with the tools to foster innovation and adaptability in a constantly evolving world. These inquiries, coupled with practical strategies for applying advanced insights, empower readers to actively engage with the challenges and opportunities presented by intricate communication systems, ultimately contributing to the broader quest for global connectivity and understanding.

Cultural Diffusion Models

Cultural diffusion is a dynamic tapestry, intricately woven over centuries of human interaction and development. Envision a world where cultural boundaries seamlessly merge into a rich blend of traditions, ideas, and innovations, continuously reshaped by the currents of global connectivity. This scenario is not merely a utopian fantasy; it arises from the complex mechanisms of cultural transmission that ripple through societal networks, much like the dynamics of wave transmission. As fresh ideas and customs permeate societies, they illuminate the paths of cultural transformation, sparking a renaissance in shared human experiences. The fusion of age-old traditions with contemporary insights creates a vibrant exchange that transcends geographical and temporal boundaries, inviting exploration into the mathematical principles that guide this cultural evolution.

At the core of this exploration is the mathematical modeling of cultural adaptation, which reveals how societies evolve over time. With technological advances accelerating cultural interchange, the patterns of convergence and divergence become more evident. Computational tools enable us to dissect these trends, offering insights into the trajectories of cultural shifts. As this narrative unfolds, it explores the transformative impact of technology on the pace and reach of cultural exchange in the digital age. This journey through cultural diffusion models not only uncovers the mechanisms of cultural growth but also highlights the potential for fostering understanding and unity in our increasingly interconnected world.

Mechanisms of Cultural Transmission Through Social Networks

The complex network of cultural transmission is heavily influenced by the dynamics present within social networks. These networks serve as channels for the dissemination of ideas, traditions, and innovations, akin to the neural pathways of a societal brain. Advanced research in network theory sheds light on how information flows through these connections, highlighting the significance of both strong and weak ties. Strong ties build trust and reinforce shared beliefs within close-knit communities, while weak ties act as bridges to diverse groups, enabling the spread of new ideas and promoting cultural exchange. This dual function of social connections emphasizes the intricate nature of cultural transmission, where the interplay between network structure and social influence determines the spread of cultural elements.

The advent of digital platforms has revolutionized cultural transmission in recent years. Social media, in particular, has accelerated the pace and expanded the reach of cultural diffusion, creating a global mosaic of interconnected communities. Algorithms that tailor content to user preferences have introduced new dynamics, sometimes reinforcing echo chambers while also providing unexpected encounters with diverse perspectives. The challenge and opportunity lie in harnessing these digital spaces to foster meaningful cultural exchanges. By strategically designing algorithms to prioritize diverse content exposure, we can enhance the richness of cultural interactions and counteract tendencies toward uniformity.

Mathematical modeling of cultural transmission offers a powerful tool to understand and predict these phenomena. Agent-based models and complex systems simulations allow researchers to explore the mechanisms underlying cultural spread, adaptation, and transformation. These models can integrate variables such as transmission rates, adoption thresholds, and social influence to replicate real-world scenarios. For example, studies on meme propagation provide insights into how cultural artifacts gain traction and evolve within digital ecosystems. By refining these models, we can better anticipate cultural shifts and develop strategies to guide their trajectories in desired directions, promoting cultural diversity and resilience.

Exploring cultural transmission through social networks also involves examining the role of influencers and opinion leaders. These individuals, often strategically positioned within networks, can either accelerate or impede the spread of cultural elements. Their endorsements or criticisms can influence public opinion and reshape cultural norms. Understanding the criteria that grant individuals such influence—whether through social capital, expertise, or charismatic authority—enables us to identify potential change agents within communities. This knowledge can be leveraged to craft initiatives that align with cultural values, ensuring sustainable and meaningful cultural evolution.

As we consider the future of cultural transmission, it is vital to address the ethical implications of technological interventions. The ability to manipulate cultural flows through network analysis and algorithmic design carries significant responsibility. Provocative questions arise: How do we balance the promotion of cultural diversity with respect for individual autonomy? What safeguards are necessary to prevent the exploitation of network dynamics for divisive purposes? By engaging with these questions, we can pave the way for a future where cultural transmission through social networks not only enriches our collective experiences but also fosters unity and understanding in an increasingly interconnected world.

Mathematical Modeling of Cultural Evolution and Adaptation

Mathematical modeling of societal evolution and adaptation unveils the complex tapestry of human communities as they transform over time. This field utilizes advanced algorithms and computational techniques to simulate and forecast the progression of societal traits. Through this perspective, societal evolution is seen as a process governed by identifiable patterns and principles, rather than mere chance. Researchers employ agent-based simulations that mimic interactions among individuals within a community, enabling the observation of emerging societal phenomena over time. These models demonstrate how small changes in individual behavior, when accumulated, can lead to significant shifts in societal identity and norms. By understanding these dynamics, we gain insights into the systematic yet organic nature of societal transformations.

The relationship between societal adaptation and environmental pressures is another crucial area of study. Mathematical models clarify how communities adjust their practices in response to environmental changes, whether ecological, technological, or social. For example, adopting sustainable practices in response to climate challenges can be analyzed through societal adaptation models. These frameworks predict how communities might alter their values and behaviors to minimize environmental impact. Researchers have developed complex equations to measure the rate of societal change and adaptation, providing insights into how communities might tackle future challenges. Anticipating these shifts allows policymakers to devise strategies aligned with emerging global trends.

Incorporating sophisticated statistical methods like Bayesian inference enables the integration of diverse data sets, enhancing the precision and depth of societal evolution models. This approach explores how societal traits spread and transform, drawing parallels with biological evolution. By applying these mathematical techniques, researchers can identify and quantify the factors driving societal divergence and convergence, offering a nuanced understanding of societal dynamics. Insights from these models are invaluable for addressing contemporary issues, such as preserving endangered languages or promoting diversity in a globalized world.

Technological progress has accelerated the pace at which societies evolve, necessitating new models to account for these rapid changes. The emergence of digital platforms and social media has dramatically altered the landscape of societal exchange, allowing information to spread rapidly. Societal evolution models now incorporate elements of network theory to understand how these technological networks affect societal transmission. By simulating information flow through digital networks, researchers can predict how societal traits might proliferate or fade in the digital era. These models highlight technology's dual role as both a catalyst for societal homogenization and a tool for preserving uniqueness.

The field of societal evolution and adaptation benefits from interdisciplinary research, drawing from anthropology, sociology, and computer science. This convergence fosters innovative perspectives that challenge traditional views and expand knowledge boundaries. By engaging with diverse theories and methodologies, researchers refine their models and enhance predictive power. As we approach new societal frontiers, these models serve as essential tools, offering glimpses into potential futures of human communities. By harnessing mathematical modeling, individuals and communities can take informed actions to guide societal evolution in ways that promote harmony and resilience in a changing world.

Impact of Technological Advancements on Cultural Exchange Rates

Technological progress has dramatically transformed the speed and nature of cultural interaction, altering how ideas and traditions spread globally. Digital communication tools, such as social media, have significantly hastened the sharing of cultural products, enabling people from various locations to engage and exchange cultural experiences at an extraordinary pace. This swift exchange fosters a more connected world, where cultural lines blur, and new hybrid cultures develop. Platforms like TikTok and Instagram serve as arenas for cultural interaction, allowing users to share and recreate content, thus promoting a lively environment for cultural evolution.

The role of technology in cultural exchange extends beyond speed to encompass reach and accessibility. Previously isolated communities can now broadcast their unique cultural views to a global audience, enriching the world's cultural landscape. Online communities offer spaces to explore, question, and appreciate cultural subtleties, enhancing understanding and appreciation of cultural diversity. Moreover, virtual and augmented reality technologies are beginning to provide immersive cultural experiences, surpassing geographical barriers and enabling profound interaction with diverse cultures.

While technological advances offer great potential, they also pose challenges that need careful handling. Although these innovations facilitate rapid cultural spread, they can also lead to cultural uniformity, where dominant cultures overshadow local traditions. This situation emphasizes the need to preserve cultural heritage in the digital era. Balancing the advantages of global cultural exchange with maintaining cultural uniqueness is a subject of ongoing debate among scholars and cultural experts. As technology advances, we must develop strategies that support cultural diversity while leveraging digital tools to enhance cultural exchange.

In understanding technology's impact on cultural spread, computational modeling offers valuable insights. Researchers use complex algorithms to simulate cultural transmission and adaptation, leading to a nuanced comprehension of how cultural traits evolve and spread over time. These models consider factors such as information flow rate, network structures, and cultural influencers. By examining these elements, scholars can predict trends of cultural convergence or divergence, aiding policymakers and cultural organizations in developing strategies for sustainable cultural exchange.

Individuals and organizations can actively enhance technology's positive effects on cultural interaction. By creating inclusive digital spaces that highlight diverse voices, users can contribute to a balanced cultural environment. Cultural institutions can use technology to curate virtual exhibitions and interactive experiences, showcasing underrepresented cultures and expanding cultural awareness. At the intersection of technology and culture, we have a significant opportunity to engage thoughtfully, ensuring the digital era enriches and preserves the global cultural landscape.

Analyzing Cultural Convergence and Divergence Patterns Using Computational Tools

Computational tools have dramatically transformed our understanding of how cultural traits spread and adapt, offering profound insights into the dynamics of cultural integration and differentiation. These sophisticated tools allow researchers to simulate and analyze intricate interactions within societal systems, uncovering hidden patterns and forces that drive cultural shifts. By modeling scenarios in digital environments, we can explore how cultural elements blend or resist integration, shaped by factors such as geographic closeness, historical connections, and social networks. These simulations reveal emergent behaviors and help forecast cultural trends, providing valuable insights for policymakers, educators, and cultural leaders.

In the study of cultural integration, computational models often utilize network analysis to track the paths through which societal ideas travel. These models illuminate the influence of key figures and nodes within social networks that facilitate cultural exchange. For example, the diffusion of technological advancements can be mapped to understand their role in promoting cultural uniformity across different regions. This method highlights the nuances of cultural adoption, adaptation, or rejection, presenting a range of outcomes rather than a binary view of cultural interaction.

On the other hand, divergence models examine how distinct cultural identities are maintained amid globalization. Computational tools identify factors that bolster cultural resilience, such as strong community bonds, language barriers, and the preservation of traditional knowledge. These models can also simulate the effects of external forces, like economic changes or policy shifts, on cultural uniqueness. By analyzing these variables, researchers gain insights into the processes of cultural differentiation and strategies to uphold cultural diversity in a rapidly evolving world.

Emerging studies investigate how digital technologies influence cultural exchange rates, as online platforms create new arenas for societal interaction. Computational analysis of online social data reveals trends in cultural preferences and discourse, shedding light on how digital communication accelerates or hinders cultural integration and differentiation. By examining digital footprints, researchers gain real-time understanding of cultural flows, offering actionable insights for those aiming to promote intercultural dialogue and collaboration online.

This examination of societal patterns through computational tools poses intriguing questions about the future of cultural identities. How will the growing complexity of digital and physical interactions reshape cultural landscapes? What influence will technology and policy have on balancing integration and differentiation? These questions encourage readers to consider their roles in the cultural ecosystems they inhabit, urging active participation in shaping a future that celebrates diversity while fostering meaningful connections across cultural boundaries. By understanding and applying these insights, readers are empowered to contribute thoughtfully to the evolving narrative of cultural change.

The dynamics of wave transmission reveal the complex pathways through which energy and information navigate various environments, connecting the tiny and vast in intriguing ways. Our exploration of subatomic wave functions has shown us how particles can display dual characteristics, challenging our understanding of reality and suggesting a deep interconnectedness at the smallest scales. Observing how information spreads highlights the ripple effect of ideas and innovations within societies, influencing networks and driving collective progress. Models of intercultural interaction further unravel the intricate web of human connection, showcasing how traditions and beliefs migrate and transform across borders, enhancing global diversity. These insights emphasize a core theme of the book: the remarkable impact of small, consistent efforts in sparking significant change. As we comprehend these principles, we are encouraged to see our actions as potential agents of transformation, prompting reflection on our role in shaping the systems around us. As we advance to the next chapter, this understanding serves as a reminder of the immense potential within seemingly minor waves of influence, urging us to harness these dynamics in our own lives.

Chapter Eight
Evolutionary Algorithms

Welcome to the fascinating world of adaptive algorithms, where the principles of survival and adaptation transcend their natural origins. Picture yourself at the boundary of a forest, observing how each creature intricately aligns with its surroundings in a dynamic interplay of life. This natural choreography reflects the profound processes mirrored in the spheres of technology and economics, where algorithms replicate survival tactics to drive progress and competition. Just as nature perfects its creations through cycles of change and selection, human innovation leverages these principles to navigate the ever-evolving landscapes of technology and market dynamics.

As we delve into this intriguing relationship, consider the similarities between genetic variations in nature and the iterative refinements in technology. These cycles of experimentation and improvement, crucial to evolution, also propel technological progress. In this chapter, we will explore how these algorithms not only spur breakthroughs but also influence market competition, guiding businesses to adapt and succeed in a constantly changing economic environment. Much like the unseen forces shaping the natural world, these patterns reveal the underlying order that dictates the success and failure of companies and technologies.

Our exploration of adaptive algorithms offers insight into the universal mechanisms connecting the individual to the collective. By understanding these patterns, we unlock the potential to enact change, whether in pioneering new technologies or navigating the intricacies of market competition. This chapter invites you to witness the significant impact of small-scale adaptations that ripple outward, creating waves of transformation across industries and societies. Embrace this journey as it sheds light on how minor actions can lead to monumental shifts, echoing the core theme of this book: the power of small actions to drive global change.

Genetic Mutation Patterns

Life's boundless diversity is driven by the fascinating process of genetic mutation, which influences everything from simple microbes to intricate organisms. Imagine a realm where tiny molecular alterations send ripples through time, shaping nature's blueprint. While often subtle at first, these genetic changes have the profound ability to direct evolutionary paths, offering a glimpse into nature's toolkit for adaptation and survival. This chapter delves into the world of these molecular catalysts, uncovering how their nuanced shifts lead to significant transformations across biological systems.

Central to this narrative is the concept of genetic drift, a quiet yet persistent force that affects mutation dynamics. Occasionally, these molecular changes reveal advantageous mutations that open up new avenues for adaptation, increasing an organism's ability to thrive in shifting environments. At the heart of these biological marvels are the molecular mechanisms that drive mutations—an unseen realm of microscopic interactions that determine the evolutionary destiny of species. As we explore further, computational models shed light on the intricate pathways shaped by mutations, offering a digital perspective to predict and understand life's evolving diversity. Through this exploration, the intricate connection between genetic mutations and evolutionary processes is revealed, providing insights into the mechanisms of innovation and survival.

The Role of Genetic Drift in Mutation Dynamics

Genetic drift, a fundamental concept in evolutionary biology, highlights the subtle yet powerful forces that influence life's diversity. Unlike natural selection, which acts through survival advantages, genetic drift quietly alters allele frequencies in populations due to random events. This random process is especially significant in small populations, where chance can cause considerable genetic changes over generations. Such changes can deeply affect a species' genetic diversity, sometimes even opposing natural selection by fixing alleles that might otherwise remain neutral or slightly harmful. Genetic drift emphasizes the unpredictability in evolution, challenging deterministic views of adaptation.

Advancements in computational biology and genomics have shed light on the complex interplay between genetic drift and mutation. High-throughput sequencing now allows scientists to observe allele frequency changes over time with unprecedented detail, offering new insights into how drift shapes evolutionary paths. Studies on isolated populations, such as those on remote islands, demonstrate how drift can cause rapid genetic differentiation. These findings highlight the importance of considering genetic drift in evaluating species' evolutionary potential, particularly in conservation biology, where genetic diversity is crucial for resilience to environmental changes.

The interaction between genetic drift and mutation adds a fascinating layer of complexity to evolution. Mutations introduce new genetic variations that drift and selection can act upon. In populations dominated by drift, the fate of new mutations is uncertain; beneficial mutations might vanish, while neutral or slightly harmful ones might become established. This randomness introduces a degree of chance to evolution, showing that random events can sometimes be as influential as adaptive pressures. Understanding this dynamic offers a more nuanced view of how species evolve and adapt, illustrating the balance between randomness and necessity.

Exploring genetic drift's practical implications, particularly in synthetic biology and genetic engineering, opens new possibilities. By applying drift principles, researchers can manage genetic diversity in engineered organisms, ensuring robust and adaptable systems. For instance, understanding drift's role can aid in predicting the long-term stability of bioengineered microbial communities. Similarly, recognizing genetic drift's impact on crop diversity can guide breeding programs to enhance yield and resistance to pests and diseases.

Considering genetic drift's influence on human evolution, especially in light of modern technological and societal changes, is thought-provoking. As humans migrate and form new, isolated communities, genetic drift could shape genetic differences among populations. This raises questions about the balance between drift and selection in human evolution and the potential for new genetic traits to emerge or disappear due to chance. Encouraging readers to explore these scenarios deepens their understanding of genetic drift and underscores the connection between small-scale genetic processes and broader evolutionary patterns, inspiring a sense of wonder at life's intricate workings.

Adaptive Significance of Beneficial Mutations

Beneficial mutations, those uncommon genetic changes that offer an advantage in an organism's environment, carry significant adaptive weight. Despite their rarity, these mutations can profoundly impact an organism's survival and reproductive abilities. In swiftly changing ecosystems, a single advantageous mutation might enhance an organism's capacity to utilize new resources or tackle emerging challenges. This potential for rapid adaptation highlights the ever-evolving nature of life, driven by the interaction of genetic diversity and environmental factors. By exploring the subtleties of these mutations, we gain insight into the forces propelling evolution forward.

Recent studies have shed light on the complex mechanisms through which beneficial mutations exert their influence. With advancements in genomic sequencing, scientists can now track these mutations at a molecular level, unveiling how tiny DNA alterations can lead to significant physical traits. For example, the development of antibiotic resistance in bacteria often originates from a few crucial mutations that modify protein structures, allowing survival in adverse conditions. Understanding these processes not only reveals nature's adaptive tactics but also equips us with knowledge to foresee and counter challenges like drug resistance.

The influence of beneficial mutations is further amplified through phenomena like genetic hitchhiking and selective sweeps. When a beneficial mutation rapidly spreads through a population, it can drag along nearby genetic variants, reshaping the genetic makeup. This demonstrates how a single advantageous mutation can affect a broader genetic context, potentially leading to new evolutionary paths. By understanding these dynamics, we can better appreciate the interconnectedness of genetic changes and their cascading effects on populations.

To apply this understanding practically, computational models have become essential tools for simulating mutation-driven evolutionary pathways. These models integrate complex variables and random elements, allowing researchers to explore hypothetical scenarios and predict evolutionary outcomes. By utilizing such models, we can simulate the potential impacts of beneficial mutations in various fields, from agriculture to medicine. These insights provide valuable guidance for strategies aimed at encouraging or mitigating evolutionary changes, depending on desired objectives.

Reflecting on the adaptive importance of beneficial mutations prompts consideration of broader evolutionary themes. How can we use this knowledge to boost resilience in ecosystems threatened by climate change? Could insights from these genetic shifts guide efforts to engineer crops more resistant to environmental stress? By asking such questions, we not only deepen our understanding of life's adaptive mechanisms but also empower ourselves to respond thoughtfully and creatively to global challenges. This exploration reminds us that the smallest genetic changes can have far-reaching effects, shaping the tapestry of life in unexpected and remarkable ways.

Molecular Mechanisms Underlying Mutation Propagation

The complex interplay of molecules within cells orchestrates the spread of genetic mutations, a fundamental phenomenon contributing to life's diversity.

On a molecular level, mutations can arise from DNA replication errors, exposure to mutagens, or the activity of mobile genetic elements such as transposons. These subtle changes can significantly alter an organism's genetic makeup. For example, point mutations involving a single nucleotide shift can drastically change a protein's function, influencing an organism's fitness and evolutionary path. Larger structural changes, including insertions or deletions, can disrupt genetic coding or create new genetic configurations. While often random, these changes are not without patterns. Understanding the molecular machinery behind these mutations provides profound insights into the evolutionary processes shaping life.

Recent advances in genomic technologies have illuminated the intricate pathways mutations follow. Techniques like CRISPR-Cas9 have not only enabled precise genetic editing but also enhanced our understanding of natural mutation processes. By observing how CRISPR-induced mutations mimic natural ones, scientists can unravel the complex balance between genetic stability and change. This research reveals that while some genome regions are highly susceptible to mutations, others remain remarkably conserved, highlighting a balance between genetic variability and preservation. Such insights are crucial in fields ranging from evolutionary biology to medicine, where understanding mutation propagation can lead to breakthroughs in disease treatment and prevention.

The study of molecular mechanisms also uncovers the influence of cellular environments on mutation rates. Factors like oxidative stress, radiation, or chemical exposure can increase mutation frequency, accelerating evolutionary processes. However, cellular repair systems, including DNA polymerases and mismatch repair pathways, work diligently to correct these errors, maintaining genomic integrity. The efficiency of these repair mechanisms can vary among organisms, contributing to the diversity in mutation rates across species. This variability underscores the importance of context in mutation propagation, where the cellular environment plays a decisive role in determining the fate of genetic alterations.

Moreover, epigenetic factors have emerged as key players in mutation dynamics. Epigenetic modifications, which affect gene expression without altering the DNA sequence, can influence mutation rates by changing the accessibility of DNA to replication and repair machinery. For instance, methylation patterns can protect against mutations by stabilizing DNA structures, yet they can also create hotspots for genetic change when disrupted. This duality offers intriguing possibilities for manipulating genetic outcomes, opening new avenues for research in gene therapy and synthetic biology. By harnessing these epigenetic insights, scientists are poised to explore new frontiers in genetic engineering, where controlled mutation propagation could revolutionize approaches to genetic diseases.

The study of mutation propagation is not merely academic; it holds practical potential for various fields. In agriculture, understanding mutation mechanisms can improve crop resilience by fostering beneficial mutations that confer resistance to pests or environmental stressors. In medicine, targeted mutation analysis could lead to personalized treatment strategies tailored to an individual's unique genetic makeup, optimizing therapeutic outcomes. By deepening our understanding of mutation propagation at the molecular level, we can unlock innovative solutions that harness genetic change's power, steering it toward beneficial outcomes that address pressing global challenges. This knowledge equips readers with the tools to envision and implement strategies that leverage the transformative potential of subtle genetic shifts.

Computational Modelling of Mutation-Induced Evolutionary Pathways

In the complex interplay of evolutionary change, computational models of mutation-driven evolutionary pathways serve as vital tools for decoding genetic adaptation's intricacies. These models, crafted using advanced algorithms, provide insights into the random nature of mutations and their cascading effects on populations over time. By simulating environments where mutations occur spontaneously and advantageous traits are favored over generations, researchers gain valuable understanding of species adaptation to shifting ecological niches. This methodology not only deepens our grasp of evolutionary biology but also offers a framework for examining how genetic variations can lead to significant evolutionary transformations.

Sophisticated computational techniques, such as Monte Carlo simulations and agent-based models, have become essential in exploring these evolutionary processes. These tools enable scientists to simulate countless scenarios, capturing the probabilistic nature of genetic variation and the subsequent selection processes. For example, in silico experiments have been crucial in forecasting organisms' adaptive potential in response to climate change by modeling genetic drift and selection pressures shaping evolutionary paths. These models help scientists anticipate potential evolutionary outcomes and inform conservation strategies for threatened species, highlighting their practical significance.

The adaptability of these models extends beyond biological systems, influencing technological innovation and market competition. By drawing parallels between genetic algorithms and market dynamics, businesses can simulate competitive environments to determine strategies that maximize growth and innovation. This computational approach can demonstrate how minor, beneficial changes ripple through complex systems, leading to substantial competitive advantages. Companies can use these insights to foster innovation cycles that mimic evolutionary processes, adapting to market shifts with agility and foresight.

Despite the promise of computational models, challenges remain in capturing the full complexity of evolutionary dynamics. The unpredictability of mutation interactions and the multifaceted nature of environmental influences necessitate ongoing refinement of models to ensure accuracy. Recent advancements in machine learning and artificial intelligence are enhancing the predictive power of these models, enabling more nuanced simulations that account for a broader range of variables. By integrating these cutting-edge technologies, researchers can refine their predictions of evolutionary pathways, further bridging the gap between theoretical understanding and practical application.

As we contemplate the implications of these computational models, a thought-provoking question arises: How can we leverage the principles of mutation-driven evolution to address contemporary global challenges? From combating antibiotic resistance to developing sustainable agricultural practices, the lessons gleaned from these models hold the promise of transformative change. By applying the logic of evolution to human endeavors, we can devise strategies that are not only resilient but also adaptive, mirroring the dynamic nature of the world we inhabit.

Technological Innovation Cycles

The story of human progress is intricately linked to the relentless march of technological advancement, a force that has continually reshaped societies and economies. From the invention of the wheel to the digital age ushered in by the internet, each wave of innovation has followed a distinct pattern of growth, upheaval, and adaptation. These cycles reflect the essence of evolutionary change, where the most effective ideas thrive, and groundbreaking developments often arise from unexpected quarters, challenging conventional wisdom. In this dynamic landscape, every technological breakthrough starts as a fledgling concept, maturing over time until it becomes an integral part of everyday life. This ongoing process, marked by the arrival of disruptive technologies, transforms markets and opens new avenues for development. By examining these patterns of change, we can trace the journey from initial concept to widespread adoption, gaining insight into the factors that drive innovations forward and those that impede them.

As new technologies evolve, they frequently encounter the transformative power of network effects and adoption curves, which significantly amplify their impact and reach. The ability of technology to connect people and ideas fosters a fertile environment for rapid growth, allowing innovations to expand swiftly and effectively. Network effects can turn an isolated technological idea into a global phenomenon, reshaping industries and societies in the process. By understanding these cycles, we can use models inspired by evolutionary algorithms to predict future trends. These models leverage principles akin to natural selection, equipping us to anticipate the paths of emerging technologies. By combining historical perspectives with advanced forecasting tools, we gain a nuanced view of how tech innovation cycles sculpt our present and carve out opportunities for the future, urging us to engage with these transformative forces thoughtfully and proactively.

Understanding the Stages of Technological Maturity

Grasping the concept of technological maturity requires an awareness of its various stages and the subtle shifts that occur between them. The journey typically begins with the nascent phase, where a technology emerges from research and development as a proof-of-concept, marked by limited functionality and application. This initial stage, akin to the early biological phases of genetic mutation, sees technologies either evolve or fade away. During this period, a flurry of experimentation occurs as researchers refine initial designs, similar to natural selection determining what innovations will endure and flourish.

As technology progresses, it transitions into the growth stage, characterized by increased adoption and market penetration. This era is often driven by substantial investments and a wave of talent focused on refining and commercializing the new technology. Here, the innovation starts to realize its potential, with network effects enhancing its value as more users and applications emerge—much like a gene that gains advantage as it spreads through a population. The rise of smartphones exemplifies this stage, evolving from niche gadgets to essential tools due to their expanding ecosystem of apps and services. This phase highlights the mutual influence between technological advancement and market forces, each propelling the other forward.

Following growth, technology reaches maturity, where expansion levels off, and the market becomes saturated. The focus shifts to optimizing efficiency and reducing costs, with improvements being incremental rather than revolutionary. This stage resembles biological homeostasis, maintaining stability through internal adjustments. The automotive industry illustrates this phase, as manufacturers continually enhance vehicle features while retaining core designs. Despite this equilibrium, opportunities for disruptive change remain, often sparked by innovations that challenge the norm.

As maturity wanes, technology faces potential obsolescence or rebirth. This pivotal moment involves choosing between reinvention or replacement. Stagnant technologies risk being overtaken by more nimble alternatives, akin to species unable to adapt to environmental shifts. Conversely, those that undergo significant transformation can experience a revival, driven by new applications and integration with emerging fields. The evolution from traditional telecommunication systems to digital networks demonstrates this renewal, where foundational infrastructure was reimagined to support progress. This stage underscores the necessity of foresight and adaptability in sustaining technological relevance.

The path through technological maturity is not a straightforward progression but a dynamic interplay of factors shaping its course. By comprehending these phases, innovators can better anticipate challenges and seize opportunities, ensuring their creations not only endure but thrive in a constantly changing environment. Consider how fields like artificial intelligence or quantum computing might redefine our current understanding of technological maturity. Such inquiries inspire strategies for capitalizing on the full potential of innovation. By aligning with these principles, individuals are equipped to navigate the intricate landscape of technological evolution, contributing meaningfully to the global narrative of change.

The Role of Disruptive Innovations in Shaping Markets

In the ever-evolving world of technology, groundbreaking advancements act as powerful forces that reshape industries and redefine market dynamics. These breakthroughs, known for their ability to disrupt established systems, introduce new solutions that significantly transform consumer habits and market frameworks. For example, digital streaming services have revolutionized the entertainment sector, making traditional cable services less critical. This shift highlights the role of transformative technologies in creating new markets while challenging existing norms. Their strength lies in addressing unmet needs or enhancing current offerings in unprecedented ways.

Examining the mechanics of disruption reveals that these breakthroughs often arise at the crossroads of different fields, where the exchange of ideas leads to innovative progress. For instance, the convergence of biotechnology and information technology has given rise to personalized medicine, which tailors healthcare to individual genetic profiles, thus revolutionizing the medical field. This fusion emphasizes the importance of interdisciplinary collaboration in generating ideas that can transform entire industries. By drawing insights from various domains, innovators are better positioned to anticipate and meet the changing demands of the market, ensuring their solutions stand out in a competitive environment.

Central to this transformative journey is the notion of creative destruction, a concept introduced by economist Joseph Schumpeter to describe the relentless cycle of innovation and obsolescence. As new technologies gain momentum, they often render older ones redundant, compelling industries to adapt or vanish. This cycle not only spurs economic growth but also fosters a culture of continuous improvement and adaptation. Organizations that embrace this mindset by encouraging experimentation and risk-taking are more likely to thrive amid disruption. Consequently, businesses are encouraged to nurture flexibility and resilience, enabling them to pivot and seize emerging opportunities.

To fully leverage disruptive advancements, it is crucial to comprehend the patterns that govern their adoption and diffusion. Theories like the innovation adoption curve offer valuable insights into how new technologies penetrate markets, emphasizing the influence of early adopters and influencers in driving widespread acceptance. By analyzing these patterns, companies can devise strategies to accelerate the adoption of their innovations, tapping into network effects and leveraging social dynamics to maximize their impact. This strategic foresight not only extends an innovation's reach but also strengthens its market position.

Looking ahead, one might consider the characteristics of future disruptive breakthroughs and their potential societal impact. As emerging technologies such as artificial intelligence and quantum computing mature, they are set to redefine industries in ways currently unimaginable. By staying alert and adaptable, individuals and organizations can position themselves to lead these changes and seize the accompanying opportunities. This proactive stance, coupled with a deep understanding of market forces, empowers stakeholders to not only respond to disruption but actively shape the future landscape.

Scaling Innovations Through Network Effects and Adoption Curves

Grasping the complex interplay between network effects and adoption curves is vital for understanding the scalability of innovations. Network effects occur when a product's or service's value increases with each new user, leading to a self-sustaining growth cycle. This phenomenon is evident in digital platforms like social media, where each additional user enhances others' experiences, resulting in rapid expansion. The interaction between network effects and adoption curves—graphs showing how quickly users embrace new technology—shapes an innovation's growth path. In today's rapid technological evolution, understanding these dynamics is crucial for innovators and strategists.

To drive adoption, innovators often design products that naturally benefit from a growing user base. For example, platform-based businesses like ride-sharing services require a substantial user base to operate efficiently. As more users join, the platform becomes more effective and appealing, attracting even more participants. The challenge is reaching the "tipping point," where growth becomes self-sustaining. This concept challenges conventional linear growth models, highlighting the need for customized strategies to boost initial user acquisition.

Recent studies emphasize the significance of timing and market readiness in leveraging network effects. Innovations gaining traction during technological or cultural shifts are more likely to be rapidly adopted. For instance, the widespread availability of smartphones enabled mobile app ecosystems to thrive due to inherent network effects. Aligning product launches with such pivotal moments allows innovators to amplify their impact by capitalizing on broader industry trends. This approach requires leaders to remain vigilant and adaptable to emerging market conditions.

Predictive models, such as those utilizing evolutionary algorithms, provide valuable tools for forecasting adoption patterns and network dynamics. These models simulate different scenarios, offering insights into potential outcomes and enabling data-driven decision-making. By leveraging these advanced techniques, organizations can predict how innovations will spread through networks, identify potential obstacles, and devise strategies to overcome them. This predictive capability is essential in navigating the complexities of modern markets, where traditional forecasting methods often fall short.

To successfully scale innovations, it's essential not only to understand network effects and adoption curves but also to actively shape them. This involves creating compelling value propositions, fostering vibrant user communities, and continuously iterating based on user feedback. Engaging with early adopters, who often serve as catalysts for broader acceptance, can accelerate this process. By nurturing these relationships and cultivating environments that encourage sharing and collaboration, innovators can significantly extend the reach and impact of their creations. As technology continues to evolve, those who master scaling through network effects will be poised to lead the future.

Predicting Future Trends with Evolutionary Algorithm Models

In the swiftly changing world of technology, evolutionary algorithm models stand out as essential tools for predicting upcoming trends. These models, inspired by natural selection and adaptation, provide a flexible framework to comprehend how technologies evolve through stages of growth, disruption, and saturation. By mimicking processes similar to biological evolution, these algorithms can analyze large datasets to discover patterns and potential paths that might otherwise go unnoticed. This capability offers a strategic edge in navigating the complexities of technological progress, enabling organizations and individuals to foresee changes and prepare for future innovations.

The true power of evolutionary algorithms lies in their capacity to model the delicate interactions between technological progress and market forces. Unlike traditional forecasting methods, which often depend on straightforward projections, these algorithms embrace the unpredictable and non-linear nature of technological ecosystems. By considering factors such as consumer preferences, regulatory shifts, and competitor strategies, they provide a comprehensive view of the innovation landscape. This broad approach facilitates more accurate predictions, particularly in industries that experience rapid change and high uncertainty, such as information technology, biotechnology, and renewable energy.

A compelling example of evolutionary algorithms at work is within the renewable energy sector. As the global focus shifts towards sustainable solutions, predicting the future development of technologies like solar panels or wind turbines becomes vital. Evolutionary algorithms can examine historical data on energy use, technological advancements, and policy changes to forecast which innovations are likely to gain momentum. This insight is crucial for investors and policymakers aiming to allocate resources effectively and promote the transition to cleaner energy sources.

Despite their robustness, it's crucial to acknowledge the limitations and potential biases of evolutionary algorithms. The accuracy of predictions largely depends on the quality of the input data; incomplete or biased data can lead to inaccurate results. Moreover, while these models offer probabilities rather than certainties, they should be seen as part of a broader decision-making toolkit. By combining algorithmic insights with human intuition and expertise, stakeholders can make more informed decisions that align with long-term strategic objectives.

When considering the future of technological advancement, it's important to address the ethical implications of predictive models. As these algorithms become more advanced, they will inevitably influence the design and implementation of new technologies. Questions about privacy, security, and equity must be considered to ensure that the benefits of innovation are shared broadly. By promoting a culture of responsible innovation, we can leverage the power of evolutionary algorithms not just to predict the future, but to shape it in ways that are inclusive and sustainable.

Market Competition Dynamics

In the competitive arena of the marketplace, businesses navigate a landscape reminiscent of adaptive systems in nature. As companies strive to outmaneuver one another, they create a complex web of interactions where small adjustments can have far-reaching effects across industries. Central to this dynamic environment are feedback loops that sustain the perpetual cycle of competition. Each decision reverberates, prompting reactions that continuously shape and propel markets. This fluidity mirrors the natural world, where survival depends on the ability to adjust and thrive amid changing conditions.

Amidst this apparent chaos, certain principles reveal the mechanisms behind major market shifts. Rapid growth can elevate a minor player to a market leader swiftly, altering the competitive scene with remarkable speed. Simultaneously, network effects magnify the reach of influential entities, enabling them to consolidate power. In this structured yet unpredictable chaos, patterns emerge that offer strategic insights into market disruption. These insights hold the potential to unlock innovative approaches for navigating the competitive landscape, echoing the book's central theme of small, strategic actions leading to significant transformations. By understanding these dynamics, one can harness the power of targeted actions to drive meaningful change on a global scale.

The Role of Feedback Loops in Competitive Markets

Feedback mechanisms are foundational in competitive markets, influencing the landscape through their dynamic interactions. These loops can magnify small advantages or setbacks, leading to substantial results. In financial markets, for example, positive feedback loops can occur as investor optimism drives stock prices higher, attracting more investors and sustaining the cycle. On the other hand, negative feedback loops can stabilize markets by counteracting deviations, such as automatic economic stabilizers that adjust tax rates or government spending in response to fluctuations. Understanding these loops helps stakeholders recognize when a market is in a self-reinforcing cycle or on the brink of a downturn.

Recent studies highlight the complexity of feedback loops in technological sectors, where rapid innovation cycles are both propelled and regulated by these loops. As companies strive to enhance their products, consumer feedback becomes a vital loop component, shaping product development and market positioning. Feedback loops in this setting support a type of evolutionary development, where products receiving positive feedback are continually improved and gain market share, much like natural selection, where only the most adaptable products succeed. Companies can use this understanding to craft more responsive and resilient strategies, ensuring competitiveness in fast-paced environments.

Feedback loops also significantly impact market competition from the perspective of behavioral economics. Human decision-making, often influenced by cognitive biases, can create feedback loops that defy traditional economic predictions. For example, herd behavior, where individuals mimic the actions of a larger group, can lead to feedback loops resulting in stock bubbles or crashes. By examining these patterns, companies and investors can better predict market movements and make informed decisions. This requires a nuanced understanding of both psychological and mathematical aspects of feedback loops, integrating insights from various fields to devise strategies that account for seemingly irrational market behaviors.

In the global marketplace, feedback loops play a crucial role in the rise and fall of market leaders. Consider network effects, where a service's value increases as more people use it, creating a powerful loop that can establish dominant market positions. Tech giants and online marketplaces leverage these loops to achieve remarkable scale and influence. However, these same loops can also lead to vulnerabilities. When negative perceptions or user dissatisfaction spread rapidly, they can quickly undermine a company's market position. Navigating these dynamics requires a strategic approach that balances growth with consumer trust and satisfaction.

As we look to the future of competitive markets, feedback loops will remain central to shaping their evolution. Companies and investors must cultivate a deep understanding of these loops to not only predict trends but also actively shape them. By fostering a culture of adaptability and responsiveness, organizations can harness feedback loops to drive sustainable growth. Envision a marketplace where feedback mechanisms are not just reactive forces but are strategically employed to anticipate and guide market trajectories. This proactive engagement with feedback can transform potential market disruptions into opportunities for innovation and leadership.

Exponential Growth and Its Impact on Market Disruption

Exponential growth carries the power to transform competitive markets, acting as a force for both innovation and disruption. This rapid and substantial increase is evident across many sectors, from technology to retail. It is not just about quick expansion; it can fundamentally alter existing structures, forcing established players to adapt or risk being left behind. Consider the impact of digital streaming services on traditional media: platforms like Netflix and Spotify have dramatically shifted consumer behavior and industry standards almost overnight, illustrating how exponential growth can overturn conventional business models.

This swift growth is often driven by technological advancements that allow businesses to scale operations efficiently. Cloud computing, for example, has enabled companies to deploy resources at unprecedented speeds, lowering barriers to entry and allowing seamless scalability. Companies like Uber and Airbnb have used this technology to disrupt the transportation and hospitality industries, demonstrating how technology can both enable and accelerate exponential growth. This often triggers a chain reaction, where initial disruptions spur further innovation across related sectors, fostering a cycle of continuous transformation.

Exponential growth also profoundly affects consumer expectations and market dynamics beyond the realm of technology. As new market entrants gain traction, they redefine value propositions, pushing competitors to innovate or face obsolescence. This shift is evident in the retail industry, where e-commerce giants like Amazon have forced traditional brick-and-mortar stores to rethink their strategies and incorporate digital elements to meet evolving consumer demands. The pressure to adapt is intense, underscoring the importance of agility and foresight in maintaining a competitive edge in rapidly changing markets.

While the benefits of exponential growth are impressive, it can also introduce volatility. Rapid expansion often brings challenges like resource constraints and potential market saturation. Companies must navigate these hurdles with strategic planning and adaptability to sustain growth over the long term. Unchecked growth can lead to instability, as seen in the dot-com bubble of the late 1990s. Understanding the nuances of exponential growth enables businesses to harness its advantages while mitigating risks.

To seize opportunities from exponential growth, businesses should cultivate a culture of innovation and experimentation. By fostering a mindset that embraces change and anticipates future trends, companies can position themselves as industry leaders. Investing in research and development helps stay ahead of technological advancements and attuned to evolving consumer preferences. By promoting adaptability and foresight, organizations can thrive amid the transformative power of exponential growth, contributing to broader market evolution and disruption.

Network Effects and Their Influence on Market Dominance

In the complex interplay of market dynamics, network effects are a powerful force that can elevate a company to market dominance. The core of network effects is the idea that a product or service becomes more valuable as more people use it. This concept not only results from user growth but also stimulates it, creating a self-perpetuating cycle of adoption. A prime example is the rise of social media platforms, where each new user adds value by offering more connections and content for others. This positive feedback loop can rapidly transform a company from unknown to dominant, as seen with platforms like Facebook and LinkedIn.

While traditional economic theories often emphasize supply and demand balance, network effects introduce additional complexity, challenging these established models. When a company effectively leverages network effects, it can reach a tipping point where growth becomes almost exponential, surpassing competitors limited by linear growth. This nonlinear trajectory can lead to a winner-takes-all scenario, where one company holds a significant market share. The consequences for competition are significant, making it difficult for new players to unseat a dominant company fortified by strong network effects.

However, the appeal of network effects is not without its challenges. Studies have shown that these effects can also pose risks, potentially leading to weaknesses. For instance, platforms heavily reliant on network effects may face scalability issues or lose user trust if network reliability is compromised. Additionally, as platforms expand, maintaining high-quality interactions is crucial. A surge of new users can dilute the experience if not managed carefully, creating a situation where growth diminishes value. Grasping these subtleties is essential for companies looking to maintain their edge in network-driven markets.

The interaction between network effects and technological progress has opened up new avenues for innovation. For instance, blockchain technology offers decentralized network effects, allowing peer-to-peer interactions without a central authority. This development poses a challenge to traditional centralized platforms, hinting at a future where network effects might be utilized in a more distributed fashion. These changes prompt businesses to reconsider strategies, focusing on community engagement and collaboration to build resilient network effects that are less prone to centralized vulnerabilities.

For businesses aiming to succeed in this environment, a strategic approach is vital. Developing a robust initial user base is key, as is using data analytics to enhance user interactions. Companies should concentrate on creating strong value propositions that not only attract users but also encourage active participation and contribution to the network. This might include fostering community spirit, offering unique incentives, or integrating complementary services to enhance the main offering. By prioritizing user experience and innovation, organizations can navigate the intricacies of network effects, transforming them from a theoretical concept into a practical tool for achieving market leadership.

Chaos Theory and Unpredictable Market Fluctuations

In the complex landscape of market dynamics, chaos theory offers a unique perspective on the unpredictable fluctuations characteristic of competitive markets. Rooted in mathematics and physics, chaos theory emphasizes the sensitivity of systems to initial conditions, demonstrating how minor variations can lead to vastly different outcomes. This principle is vividly illustrated in the financial sector, where subtle changes in investor sentiment or geopolitical events can trigger significant market swings. Such volatility highlights the importance of discerning patterns that might initially appear inscrutable but reveal a deeper order when viewed through the lens of chaos theory.

Consider the stock market, where traders and analysts contend with the unpredictability of price movements. Traditional models strive to forecast trends using historical data, yet chaos theory posits that these systems are fundamentally nonlinear and sensitive to initial conditions, rendering precise predictions nearly impossible. This understanding has spurred the creation of advanced algorithms that incorporate chaotic dynamics to better navigate the market's intricacies. By embracing market unpredictability, investors can devise strategies that are more adaptable to sudden changes, transforming volatility into opportunity rather than perceiving it as a threat.

Cutting-edge research in chaos theory has also illuminated the concept of attractors, which are states toward which a system naturally gravitates. In market competition, attractors can represent dominant players or niches that consistently draw resources and attention. This insight provides a strategic edge, enabling companies to identify and capitalize on potential attractors within their industries. By understanding these attractors, businesses can position themselves advantageously, anticipating shifts in market dynamics and tailoring their strategies accordingly, thus ensuring long-term competitiveness.

Chaos theory also challenges the conventional wisdom of linear cause-and-effect relationships in market analysis. It encourages a more nuanced exploration of the interconnected elements that drive market behavior. For example, a minor technological advancement in one sector can trigger a cascade of changes across multiple industries. By adopting a holistic view, market participants can better appreciate the complex interplay of factors influencing market outcomes, leading to more informed decision-making processes. This broader perspective enhances strategic agility, enabling stakeholders to adapt swiftly to emerging trends and unforeseen disruptions.

To effectively navigate the chaotic nature of market fluctuations, it is crucial for businesses and investors to cultivate a mindset that embraces uncertainty as a catalyst for innovation. By fostering a culture of continuous learning and adaptation, organizations can harness chaos as a driving force for creative problem-solving and strategic growth. Encouraging experimentation and flexibility allows for the development of agile responses to volatile market conditions. This proactive approach not only mitigates risks but also uncovers hidden opportunities within the chaos, empowering businesses to thrive in an ever-evolving competitive landscape.

Evolutionary algorithms vividly illustrate the fascinating connection between the adaptability found in nature and the progress of human innovation. Observing genetic mutations, we see that small, incremental changes can yield significant transformations. This concept extends beyond biology, influencing the trajectory of technological advancement. The iterative cycle of innovation mirrors these natural processes, where trials and errors ignite the evolution of ideas, culminating in breakthroughs that redefine industries. In parallel, the dynamics of market competition reflect the principle of survival, where adaptability and strategic insight pave the way to success.

These insights highlight how individual actions can ripple out to produce significant global outcomes, reinforcing the core theme of this book: small patterns can drive extensive change. As we deepen our understanding of these algorithms, we gain the tools to leverage them effectively, fostering growth and innovation in our environments. Readers are encouraged to contemplate how these principles, when applied, can spark positive change in their own spheres. The journey continues, promising further exploration of evolving patterns that enhance our understanding, equipping readers to contribute meaningfully to the ever-changing landscape of global transformation.

Chapter Nine
Chaos And Predictability

When you were a child, you might have been captivated by a raindrop's sudden splash on the pavement, scattering in a dance of tiny droplets. This ordinary moment offers a peek into the intriguing interplay of disorder and order—an intricate ballet that orchestrates much of our world. Imagine the excitement of learning that the same forces guiding that raindrop's unpredictable journey also shape the swirling patterns of weather systems, the erratic fluctuations of financial markets, and the vibrant cycles of ecosystems. This chapter invites you to explore this domain where random events uncover the hidden structure beneath, revealing the delicate balance between chaos and harmony.

Consider the butterfly effect, a concept suggesting that a butterfly flapping its wings in Brazil could trigger a tornado in Texas. This idea exemplifies how small shifts can lead to significant, unforeseen outcomes, echoing across various systems. As we delve into the complex fabric of chaos, you'll encounter the capricious nature of weather, where slight atmospheric changes can transform entire forecasts. Similarly, the stock market, often seen as an enigmatic entity, is influenced by myriad factors, each contributing to its volatile character. Moving further, the subtle interplay of population changes highlights how minor alterations in birth rates or resources can impact entire species, sometimes dramatically.

These explorations not only deepen our understanding of disorder but also illuminate the thread of order within it. By examining the forces that govern these phenomena, you'll discover the mathematical elegance at their core. This journey offers more than just understanding; it empowers you to apply these principles, fostering positive change in the systems you influence. As you navigate through this chapter, you'll gain a renewed appreciation for the complex designs shaping our world and the potential to harness them in your quest for transformation.

Weather Pattern Formation

Step into the captivating world of atmospheric phenomena, where the intricate dance of weather systems unfolds a tapestry woven with both chaos and order. Here, even the slightest change in air pressure can herald vast shifts, transforming gentle breezes into powerful forces. The atmosphere, a vast and complex web of air and moisture, crafts mesmerizing and mysterious designs. Every fluctuation in pressure acts like a note in this grand symphony, guiding the creation of clouds, the arrival of storms, and the peace of clear skies. These movements, though seemingly random, follow underlying principles that echo through the realms of mathematical chaos and predictability, inviting us to uncover their hidden harmonies.

As we expand our view, the influence of oceanic currents becomes apparent, shaping global climate patterns with their mighty flow. Driven by differences in temperature and salinity, these currents form a network that distributes heat and nutrients, affecting weather patterns far beyond their watery paths. The sun, with its ever-present warmth, layers additional complexity onto our weather narrative. Solar radiation, with its cyclical patterns, acts as both creator and disruptor, painting the skies with warmth and altering weather variability with subtle shifts. The interplay of these elements creates a captivating dynamic, where jet streams and cyclonic activities spring to life, orchestrating the fascinating and unpredictable symphony of weather. Each element reveals a piece of this elaborate puzzle, inviting us to appreciate the interconnected forces shaping our world, illustrating how the smallest changes can resonate across the vast expanse of our planet's atmosphere.

The Role of Atmospheric Pressure in Weather Systems

Atmospheric pressure acts as an unseen force, shaping the various weather systems that move across our planet. Its changes create a web of interactions that drive wind currents, mold cloud formations, and affect rainfall. Essentially, atmospheric pressure is the force exerted by the air above a specific location, with high-pressure areas typically linked to clear skies and low-pressure zones often bringing clouds and storms. The movement of these pressure zones around the globe is responsible for the different weather patterns seen in various regions. Modern meteorological research has highlighted the complex behavior of pressure systems, revealing their influence on phenomena like cyclones and anticyclones, which can significantly change weather conditions over large areas.

A captivating aspect of atmospheric pressure is its interaction with Earth's rotation, known as the Coriolis effect. This effect causes air masses to deflect, affecting wind direction and speed. In the Northern Hemisphere, air tends to move clockwise around high-pressure areas and counterclockwise around low-pressure zones, with the opposite occurring in the Southern Hemisphere. Understanding this effect is crucial for weather forecasting, as it helps meteorologists predict storm paths. Advanced models now include this factor, resulting in more accurate forecasts and allowing communities to better prepare for weather-related events.

Exploring how atmospheric pressure relates to weather systems also provides insights into climate change. As global temperatures rise, pressure systems are expected to shift in distribution and strength, potentially leading to more extreme weather events. For instance, high-pressure systems might become more persistent, causing longer heatwaves, while stronger low-pressure systems could result in more frequent and intense storms. Cutting-edge research is focused on understanding these potential changes, using advanced computational models to simulate future scenarios and assess their impact on global weather patterns.

Atmospheric pressure is also crucial in the phenomenon known as atmospheric rivers—narrow streams of concentrated moisture in the atmosphere. These rivers can carry vast amounts of water vapor across great distances, resulting in significant rainfall when they make landfall. Understanding the pressure systems that guide these atmospheric rivers is vital for predicting their landfall and mitigating flood risks in affected regions. Recent studies emphasize the importance of monitoring pressure changes to anticipate such events, offering a proactive approach to disaster preparation and response.

For those interested in these complex processes, a deeper understanding of atmospheric pressure's role in weather systems offers valuable insights. By observing barometric changes, individuals can make informed decisions about agricultural practices, disaster planning, and even recreational activities. Those in aviation or maritime navigation can use knowledge of pressure patterns to optimize routes and enhance safety. As we continue to explore the intricacies of atmospheric pressure, we empower ourselves to leverage its impact, transforming our understanding of weather from passive observation into an active tool for navigating the complexities of our environment.

Oceanic Currents and Their Impact on Climate Patterns

Ocean currents are the unseen forces shaping our planet's climate, intricately influencing weather patterns. These massive flows of seawater, driven by winds, salinity differences, and temperature variations, distribute heat and nutrients globally. The Gulf Stream, for instance, carries warm waters from the Gulf of Mexico across the Atlantic to Western Europe, granting the region a milder climate than its latitude might suggest. This complex interaction between ocean and atmosphere is crucial for regulating weather systems, impacting phenomena from South Asian monsoons to African droughts. By grasping the mechanics of these currents, we can enhance our ability to predict and adapt to the climatic changes they bring.

The field of oceanography continually reveals the intricacies of these currents. Recent research highlights the thermohaline circulation, often called the "global conveyor belt," as essential for maintaining climate balance. This global ocean current is driven by changes in water temperature and salinity, causing denser, colder water to sink and flow along the ocean floor, while warmer, lighter water stays near the surface. Disruptions to this circulation, possibly due to melting polar ice reducing salinity, could have significant impacts, such as altering rainfall patterns and intensifying extreme weather events. The potential for such disruptions emphasizes the need to monitor oceanic changes and incorporate these findings into climate models.

Exploring the relationship between ocean currents and climate patterns also illuminates feedback mechanisms that can amplify or diminish climatic changes. For example, the El Niño-Southern Oscillation (ENSO) shows how shifts in oceanic and atmospheric conditions can lead to global weather anomalies. During an El Niño event, warm water in the Pacific Ocean disrupts atmospheric circulation, causing heavy rainfall in some areas and droughts in others. These feedback loops underscore the interconnectedness of Earth's systems, where changes in one component, such as ocean currents, can ripple through the entire climate system, affecting agriculture, water resources, and even socio-economic stability.

Given these complexities, innovative technologies and methods are emerging to better understand and predict the impact of ocean currents on climate. Satellite-based remote sensing and autonomous underwater vehicles now provide unprecedented data on ocean temperatures, salinity, and currents. Machine learning algorithms analyze this data, identifying patterns and anomalies that might otherwise remain hidden. These advancements not only enhance our predictive capabilities but also inform policy decisions and resource management strategies. By leveraging modern technology, scientists and policymakers can collaborate to mitigate the adverse effects of climate change and harness these natural forces for sustainable development.

As we delve deeper into the profound influence of ocean currents on climate, a vital question arises: how can we use this knowledge to build resilience in communities most affected by climate variability? Understanding oceanic influences enables more accurate forecasting, which can inform agricultural planning, disaster preparedness, and water management. Encouraging a holistic perspective that integrates scientific insights with local knowledge can empower communities to adapt to changing conditions. By fostering collaboration across disciplines and borders, we can transform our understanding of ocean currents into actionable strategies that promote environmental stewardship and global well-being.

The Influence of Solar Radiation on Weather Variability

Solar radiation, the primary catalyst of weather variability, orchestrates a complex interplay of atmospheric events. This energy, radiating from the sun, actively influences Earth's climate systems rather than serving as a static backdrop. Even minor fluctuations in solar output can trigger significant shifts in weather patterns. For example, the solar cycles that occur roughly every 11 years impact the stratosphere's temperature, which in turn modifies the jet stream's course. Such changes can result in unexpected weather, such as extended droughts or sudden cold spells, highlighting the substantial effect of solar variations on Earth's weather systems.

The interaction between solar radiation and Earth's albedo further illustrates its impact on weather dynamics. Albedo, the measure of Earth's reflectivity, determines how much solar energy is absorbed or reflected back into space. Human activities, like deforestation or urban development, can alter local albedo and thereby affect local climate conditions. Urban areas, with their dark surfaces like asphalt, typically have lower albedo, leading to heat islands that can intensify local weather events, including thunderstorms. This emphasizes the importance of considering both natural and human influences when assessing solar radiation's role in weather variability.

Advancements in satellite technology have significantly expanded our understanding of solar radiation's impact on weather systems. Tools like the Solar and Heliospheric Observatory (SOHO) continuously monitor solar activity, providing insights into how solar flares and coronal mass ejections affect Earth's magnetosphere and climate. These interactions can disrupt communications and power grids, illustrating the broader implications of solar variability. By utilizing these technological advancements, researchers can create more accurate predictive models, enhancing our ability to foresee and mitigate extreme weather events.

The multifaceted relationship between solar radiation and weather variability also highlights the potential for renewable energy solutions to mitigate climate change. Harnessing solar energy through photovoltaic systems offers a sustainable power source, reducing reliance on fossil fuels and decreasing greenhouse gas emissions. Transitioning to renewable energy can help stabilize atmospheric conditions by minimizing human impacts, ultimately contributing to a more balanced and predictable climate system. Encouraging the adoption of solar technologies thus presents a practical approach to addressing both energy needs and weather variability.

Reflecting on the broader implications of solar radiation on weather patterns invites contemplation on humanity's role in shaping its environment. How can societies better integrate solar energy into their infrastructure to build resilience against weather fluctuations? What innovative policies could incentivize the reduction of urban heat islands while promoting green spaces that enhance albedo? These questions challenge us to engage critically with the knowledge presented, underscoring the opportunity for individuals and communities to enact meaningful change. By embracing a holistic understanding of solar radiation's influence, we can cultivate a future where small-scale actions contribute to global climate stability.

Interactions Between Jet Streams and Cyclonic Activities

Jet streams, those swift bands of wind high above our planet, are crucial in influencing cyclonic behaviors. These atmospheric expressways, coursing thousands of kilometers over our heads, do more than just transport air—they actively shape weather systems. Their interaction with cyclones can either bolster or weaken these powerful storms. When a jet stream aligns with a growing cyclone, it can boost the storm's potency by offering an escape route for air, a phenomenon known as jet stream-induced intensification. This process has become a cornerstone of recent meteorological research, highlighting the intricate balance of atmospheric interactions.

Beyond storm enhancement, jet streams significantly affect the paths of cyclonic systems, impacting whether they head toward populated regions or remain over open seas. For example, the North Atlantic Oscillation, a climatic event influenced by jet streams, can alter hurricane trajectories, sometimes diverting them from the U.S. eastern seaboard. This illustrates the profound influence of upper-atmosphere movements on surface weather events, an area that continues to intrigue scientists eager to decode atmospheric complexities.

In the context of climate change, the behavior of jet streams is under increasing scrutiny. As global temperatures climb, these air currents are shifting in unexpected ways, with possible consequences for cyclonic activity. A warming Arctic, for instance, is thought to weaken the polar jet stream, causing it to wander and potentially result in extended weather patterns. This "jet stream waviness" can lead to prolonged periods of storms or calm, affecting the frequency and severity of cyclonic events. Understanding these changes is critical for forecasting future weather patterns and preparing for their impacts on human societies and ecosystems.

The study of jet streams and their interaction with cyclones extends beyond theoretical exploration to practical applications. Meteorologists utilize sophisticated computational models to simulate these interactions with remarkable accuracy, enhancing weather forecast precision. These models consider numerous factors, from the upper-level divergence of air to the release of latent heat within storms, offering a comprehensive view of how jet streams influence cyclonic activity. By refining these models, scientists aim to deliver more dependable predictions, aiding in disaster preparedness and risk mitigation globally.

These atmospheric complexities prompt us to consider the broader implications of such phenomena on our understanding of weather systems. How might future technological and research advancements further illuminate the role of jet streams in cyclonic activities? Could interdisciplinary approaches, integrating insights from oceanography and climatology, unlock new perspectives on atmospheric behavior? As we ponder these questions, the vast, interconnected web of atmospheric science stands as a testament to the profound complexity and beauty of our planet's weather systems, inviting both seasoned researchers and curious minds to explore its ever-evolving mysteries.

Stock Market Fluctuations

The stock market's fluctuations are a captivating spectacle, appearing unpredictable yet governed by intricate patterns reminiscent of nature's own chaos and order. Much like weather phenomena or the ebb and flow of ecological populations, the market strikes a delicate balance between stability and turbulence, where minor shifts can lead to monumental impacts. This delicate interplay invites us to delve deeper into the myriad factors influencing market behavior. From the subtle sway of investor moods to the resonant echoes of past market volatility, every element plays a role in the complex choreography of financial trends. As we journey through this dynamic landscape, we begin to unravel the mathematical principles that weave through its movements, revealing an underlying structure within the apparent disorder.

In this whirlwind of economic activity, the stock market provides fertile ground for exploring the interplay of chaos and order. Historical patterns of volatility reveal recurring themes, while behavioral feedback loops underscore the significant power of sentiment in driving market shifts. Fractal geometry offers a lens to view the repetitive nature of market trends, echoing patterns found in natural systems. Additionally, the realm of quantum finance introduces a fascinating layer of uncertainty, challenging conventional notions of predictability in trading. These interconnected threads form a rich tapestry of insights, inviting readers to appreciate the intricate beauty and complexity of the financial market's ever-evolving dance.

Analyzing Historical Market Volatility Patterns

The allure of market turbulence captivates both analysts and investors, presenting both hurdles and prospects. To grasp the historical ebb and flow of such volatility, one must delve into the intricate and unpredictable dynamics of financial systems. By scrutinizing past variations, we can discern recurring motifs and glean insights into future tendencies. Historical records indicate that market fluctuations often exhibit cyclical trends shaped by economic signals, geopolitical events, and technological progress. These cycles, however, are not regular, adding layers of complexity to forecasting. Studying these cycles fosters a deeper understanding of the market's multifaceted nature and lays the groundwork for crafting more refined analytical frameworks.

In this context, sentiment analysis stands out as a vital instrument, deciphering the collective mindset of market players and its influence on volatility. Behavioral feedback mechanisms, where traders' emotions and actions mold market outcomes, can create a self-fulfilling prophecy that magnifies price shifts. Grasping these loops demands a nuanced perspective that incorporates both quantitative data and qualitative insights into market psychology. For instance, the irrational enthusiasm of a bull market can drive unsustainable asset price surges, which eventually correct themselves, often with marked volatility. By blending sentiment analysis with traditional technical and fundamental evaluations, investors can more effectively anticipate changes in market trends and volatility.

Fractal geometry offers yet another fascinating perspective on market trends, emphasizing the self-similar nature of market movements across varying timeframes. Much like fractals, which display repeating patterns at progressively smaller scales, financial markets can exhibit similar structures within their fluctuations. This fractal characteristic suggests that market trends might not be entirely random but instead guided by underlying principles of self-similarity. By recognizing these structures, traders and analysts can devise strategies that exploit the predictability of certain market movements, even amidst apparent disorder. The use of fractal analysis can pave the way for innovative trading models, offering a fresh take on market behavior.

Quantum finance introduces an innovative approach by applying quantum mechanics principles to the uncertainty inherent in trading. Market volatility resembles the uncertainty principle, where measuring one variable precisely impacts the certainty of another. In trading, this manifests as the challenge of accurately predicting both price and timing. Quantum finance posits that embracing uncertainty, rather than trying to eliminate it, can yield more resilient investment strategies. This paradigm shift encourages traders to concentrate on probabilistic outcomes and risk management, promoting a more adaptive stance toward market fluctuations.

These varied viewpoints on market volatility highlight the necessity of a comprehensive approach to understanding and navigating financial landscapes. By merging historical patterns, sentiment analysis, fractal geometry, and quantum finance, market participants can hone their strategies and adapt to shifting conditions. Encouraging a mindset of curiosity and continuous learning, these insights empower readers to engage with financial markets more informatively and proactively. As they apply these concepts, they contribute to the broader discussion on market dynamics, fostering innovation and resilience in the face of uncertainty.

The Role of Sentiment and Behavioral Feedback Loops

In the complex realm of stock markets, the influence of sentiment and behavioral feedback loops is fundamental to understanding market behavior. These loops are primarily fueled by the collective mindset of investors, whose emotions and perceptions can drastically alter market trends. When investors feel optimistic, buying activity can increase, driving prices higher. This surge in confidence can form a self-perpetuating cycle, as rising prices draw in more investors hoping to benefit from the gains, further amplifying the upward trend. Conversely, fear can trigger sell-offs, leading to a downward spiral. This underscores the significant role of sentiment in market volatility, where emotions often override rational analysis.

Recent studies provide new perspectives on these feedback loops, highlighting the complexity of investor sentiment. Today, machine learning and sentiment analysis tools sift through vast data from news, social media, and trading platforms to assess investor mood. Advanced analytics uncover patterns that were once hidden, offering a deeper understanding of how sentiment changes over time. This challenges the traditional efficient market hypothesis, suggesting that markets are not always logical and that psychological factors can create predictable volatility patterns. By identifying these patterns, traders and analysts can better foresee market movements, allowing them to strategically position themselves.

The idea of behavioral feedback loops also aligns with George Soros's theory of reflexivity. Reflexivity posits that market participants' perceptions can influence market fundamentals, which then alter perceptions, forming a continuous loop. This idea highlights the dynamic relationship between belief and reality in financial markets. For instance, if a large group of investors believes a stock is undervalued, their collective actions might increase its price, confirming their initial belief. Reflexivity offers a framework to understand how self-fulfilling prophecies occur in markets, underscoring sentiment as both a cause and consequence of market changes.

To leverage sentiment, investors might adopt a contrarian strategy, taking advantage of the cyclical nature of emotions. By recognizing extreme optimism or pessimism, astute investors can position themselves to benefit from eventual market corrections. For example, during times of excessive euphoria, a contrarian might reduce their exposure, anticipating a market correction. Conversely, in periods of widespread fear, they could increase their investments, expecting a rebound. This approach requires a deep understanding of market psychology and the ability to remain detached from dominant sentiments, focusing instead on long-term fundamentals.

As sentiment analysis continues to advance, innovative strategies are emerging to navigate the intricate landscape of behavioral feedback loops. Investors increasingly rely on technology-driven solutions, such as AI and big data analytics, for real-time insights into market sentiment. These tools help them make informed choices, reducing dependence on intuition and enabling more strategic approaches. By combining these cutting-edge methods with traditional analysis, investors can gain a more comprehensive understanding of market behaviors, positioning themselves to succeed in an environment where sentiment and actions are tightly intertwined.

Fractal Geometry and the Self-Similarity of Market Trends

Fractal geometry unveils an enchanting mosaic within the stock market, revealing designs that echo across scales with captivating self-similarity. Introduced by Benoit Mandelbrot, this concept suggests that market trends are not random fluctuations but possess intricate structures that repeat across different time frames. Just as a jagged coastline appears complex yet orderly from any viewpoint, stock market charts exhibit a fractal nature where designs recur whether one examines hourly, daily, or annual data. This repetition is not merely visual but mathematical, as market movements can often be modeled using fractal mathematics, providing traders and analysts with a sophisticated lens to interpret volatility and trend formation.

Delving deeper into fractal analysis equips investors with tools to anticipate market behavior by identifying recurring motifs. For instance, the Elliott Wave Theory, grounded in fractal concepts, proposes that market trends unfold in predictable wave patterns. These patterns are visible in both bullish and bearish markets, where each wave is part of a larger fractal structure. By recognizing these waves, traders can potentially forecast future price movements, gaining a strategic advantage. Acknowledging self-similar designs within the market arms investors with a nuanced understanding, allowing them to navigate the turbulent seas of trading with increased foresight and precision.

The integration of fractal geometry with modern computational techniques heralds a new era in financial analysis. Advanced algorithms, capable of identifying fractal dimensions within market data, offer unprecedented insights into market behaviors. Machine learning models, trained on these fractal patterns, can predict market movements with a level of accuracy previously considered unattainable. This synergy between fractal analysis and artificial intelligence is transforming traders' approach to market prediction, presenting opportunities to rethink conventional strategies and embrace innovative methods that harness the power of fractal mathematics.

Fractal geometry also challenges the conventional wisdom of market efficiency, suggesting that markets are not always rational or linear. The fractal nature of market trends implies that underlying processes may be governed by complex, non-linear dynamics. This perspective encourages traders and analysts to consider factors beyond traditional financial indicators, such as psychological and sociological influences that may manifest in fractal patterns. By expanding their analytical framework to include these dimensions, investors can develop a more comprehensive understanding of market behavior, enabling more informed decisions in the face of uncertainty.

Embracing this fractal perspective invites investors to reframe their approach to trading, highlighting the importance of adaptability and mindfulness in strategy formulation. By cultivating an awareness of market fractals, traders can refine their skills in identifying designs and develop more resilient strategies to withstand the unpredictable nature of financial markets. This approach not only enhances their ability to capitalize on market opportunities but also fosters a deeper appreciation for the intricate beauty of market dynamics, inspiring a sense of curiosity and innovation in their trading endeavors.

Quantum Finance and the Uncertainty Principle in Trading

Quantum finance explores the fascinating overlap between physics and financial markets, offering deep insights into the inherent unpredictability of trading. Drawing from the uncertainty principle in quantum mechanics, which suggests that certain physical properties like position and momentum cannot be precisely known at the same time, quantum finance uses this concept metaphorically to address the stock market's complexity. Traders face a similar challenge: the more accurately one market variable is predicted, the more uncertain other factors become. This requires a delicate balance between precision and uncertainty, urging traders to adapt their strategies to the market's volatile nature.

By integrating quantum theories into financial models, traditional deterministic approaches are challenged, encouraging a shift toward probabilistic frameworks that better reflect the chaotic nature of markets. This change enhances our understanding of market fluctuations and investor behavior. Recent developments in quantum finance highlight the significance of interconnected market variables, where a change in one area can instantly affect another. This insight pushes traders to consider the broader effects of their decisions, promoting a holistic approach to market analysis and strategy formulation.

Fractal geometry, with its recurring patterns at various scales, complements the quantum perspective by illustrating how small-scale market behaviors can mirror larger trends. This fractal nature suggests that patterns seen in minute-by-minute trading could provide insights into broader movements. By studying these fractal structures, investors gain a deeper understanding of market trends, helping them identify opportunities and risks that might not be evident through conventional analysis. Recognizing recurring market cycles and patterns across different time frames gives a strategic edge in both short-term trading and long-term investment planning.

As quantum finance progresses, new tools and methods emerge, enabling traders to leverage principles of uncertainty and interconnectedness. Quantum computing, for instance, has the potential to transform trading algorithms by processing vast data sets at unprecedented speeds. This ability allows for real-time assessment of complex market scenarios, giving traders a competitive advantage in adapting swiftly to changes. The integration of quantum-inspired algorithms into trading platforms marks a new frontier in financial technology, promising enhanced predictive accuracy and strategic adaptability.

This convergence of quantum mechanics and finance raises thought-provoking questions: How can traders reconcile market unpredictability with the need for actionable strategies? What ethical considerations arise when using powerful quantum computing resources in trading? By challenging conventional thinking and exploring these questions, investors and researchers are encouraged to critically assess the future of finance. This exploration not only deepens our understanding of market dynamics but also inspires proactive use of scientific principles for positive change in the financial sector.

Population Dynamics

Imagine waking up to a world buzzing with life, where every creature engages in a delicate interplay of growth and balance. From the minuscule microorganisms in the earth to the majestic mammals of the savannah, each species follows a rhythm shaped by a complex web of interactions. The science of population dynamics unveils the hidden designs that dictate the rise and fall of life, highlighting nature's astonishing ability to self-regulate, adapt, and sometimes descend into disorder. Here lies a captivating contradiction: the certainty of growth is entwined with the unpredictability of environmental challenges and interspecies relationships. As we delve into this subject, we discover how minor shifts—like an increase in predators or a slight change in climate—can resonate through ecosystems, sparking transformations that reverberate across time and space.

Central to these interactions are mathematical models that strive to encapsulate the essence of growth and regulation. These models expose a realm of feedback mechanisms where stability and unpredictability coexist. In predator-prey interactions, chaos theory illustrates how small changes can lead to significant shifts in population sizes. Additionally, synchronization phenomena reveal how individual actions can align to form collective behaviors, highlighting the power of connections. As we explore these themes, the elaborate beauty of population dynamics becomes evident, offering insights into the universal principles that shape the intricate fabric of life. Each discovery serves as a reminder of the profound implications these principles hold for understanding not only the natural world but also the systems we build and live within.

Mathematical Models of Population Growth

Population growth is intricately woven with mathematical complexities, intertwining birth rates, death rates, and environmental influences to form dynamic systems that resist straightforward explanations. At the heart of this complexity lie mathematical models, which provide essential frameworks for predicting and analyzing how populations change over time. The exponential and logistic growth models are particularly fundamental, offering insights into population expansion in both ideal and resource-constrained environments. The exponential model, often demonstrated by the unchecked growth of bacteria in a petri dish, assumes boundless resources, resulting in rapid increase. Conversely, the logistic model introduces the concept of carrying capacity, where resource limitations cap growth, creating an S-shaped curve that better mirrors reality.

Advancements in computational power and data analytics have led to the development of more sophisticated models that incorporate randomness and nonlinear dynamics. These models acknowledge that real populations do not exist in isolation; instead, they are influenced by random environmental changes and complex interactions with other species. Agent-based models, for instance, simulate individual entities within a population, each with specific behavioral rules, to predict larger phenomena. These models have been crucial for studying disease spread, where human behavior and social networks significantly affect outcomes. By embracing the inherent randomness and complexity of population systems, these models offer deeper insights into growth dynamics and potential tipping points.

In cutting-edge research, the fusion of machine learning with population models is particularly promising. Machine learning excels at recognizing patterns, allowing the identification of subtle trends and anomalies that traditional models may miss. By training algorithms on vast datasets, researchers can uncover hidden correlations and predict future population changes with remarkable accuracy. This is especially valuable in conservation biology, where anticipating how environmental changes affect endangered species is critical. Machine learning models can pinpoint key factors influencing population viability, aiding in the creation of targeted conservation strategies that ensure the survival of vulnerable species.

Diverse perspectives enrich the discussion on population dynamics, challenging conventional narratives and encouraging innovative solutions. Some scientists emphasize the inclusion of cultural and social factors in population models, highlighting how human populations are uniquely affected by beliefs, traditions, and policies. Integrating these dimensions can provide insights into phenomena like urbanization, migration, and demographic transitions. Moreover, there are differing opinions on technology's role in population growth, with some viewing it as a catalyst for sustainable development and others warning of its potential to increase inequalities and resource exploitation. Exploring these varied viewpoints helps develop more comprehensive models that capture the multifaceted nature of population dynamics.

Engaging with the complexities of population growth encourages readers to consider how mathematical models can address real-world challenges. A thought-provoking example might involve a city planner using these models to predict urban growth and design infrastructure that meets future needs. By analyzing factors like migration patterns, birth rates, and economic trends, planners can make informed decisions balancing development with sustainability. On a practical level, readers might explore how population models influence personal decision-making, such as assessing lifestyle choices' impact on community resources. By fostering a proactive mindset, this exploration empowers individuals to contribute to positive change through informed, mindful actions.

The Role of Feedback Loops in Population Regulation

Feedback loops are essential in controlling population dynamics, where the relationship between population size and environmental conditions creates intricate interactions. These loops can be either reinforcing or balancing, each influencing the population's path differently. Reinforcing loops might trigger rapid growth when resources are plentiful, leading to a swift rise in population numbers. On the other hand, balancing loops often serve to stabilize growth, limiting expansion when resources dwindle or when the population hits the environment's capacity limits. This delicate balance is evident in ecosystems where the ebb and flow of predator-prey dynamics naturally sustain ecological stability.

A prime example of balancing feedback in population control is the classic predator-prey model. As prey numbers rise, they offer more sustenance for predators, which then reproduce more successfully, increasing their count. However, as predator numbers climb, they put more pressure on the prey population, eventually causing a decline in prey. This reduction subsequently leads to fewer predators due to limited food, establishing a cyclical rhythm. The Lotka-Volterra equations mathematically represent this dynamic, offering profound insights into how feedback loops can maintain order amidst seeming disorder.

In recent times, researchers have examined how human actions disrupt these natural feedback systems, often with unforeseen outcomes. Farming methods, urbanization, and deforestation change habitats and resource availability, causing imbalances in population dynamics. For instance, reducing natural predators through hunting or habitat loss can lead to unchecked prey populations, which may then surpass the environment's capacity to support them. Such unchecked growth can deplete resources and cause population crashes, highlighting the fragile balance maintained by feedback loops in nature.

Advancements in computational modeling have deepened our understanding of feedback loops' effects on population regulation. Tools like agent-based models allow researchers to simulate complex ecosystem interactions, providing insights into how changes in one factor can ripple through the system. These models help forecast the outcomes of different scenarios, such as introducing a new species or removing a key predator. By understanding these dynamics, conservationists can devise strategies to promote biodiversity and ecosystem resilience, ensuring human interventions are in harmony with nature's intrinsic regulatory mechanisms.

Exploring these ideas prompts a deeper consideration of our role in preserving ecological balance. As caretakers of the planet, how can we leverage feedback loops to encourage sustainable practices and mitigate adverse effects? By examining these natural processes, we gain a clearer understanding of life's interconnectedness and the potential to guide population dynamics toward harmonious coexistence. This knowledge empowers us to make changes that respect and preserve the intricate dance of life, inviting readers to appreciate not only the mathematical elegance of feedback loops but also their practical application in fostering a sustainable future.

Chaos Theory in Predator-Prey Dynamics

In the complex web of ecosystems, the interplay between predators and prey offers a prime example of chaos theory in action. These interactions, though seemingly random, are governed by mathematical principles. Central to this complexity is the sensitivity to initial conditions—a key feature of chaos theory. Even a slight shift in predator or prey numbers can cause significant changes over time, creating a dynamic balance that appears chaotic but is actually deterministic. This understanding challenges the conventional view of randomness, revealing an underlying structured chaos that can be modeled and partially predicted.

Recent studies have shed light on the chaotic nature of predator-prey relationships, providing new insights into population cycles that defy linear expectations. The well-known case of the Canadian lynx and snowshoe hare illustrates how chaos manifests in natural settings. Advanced computational models have shown that these cycles are not solely the result of external environmental influences but are also deeply affected by the internal dynamics of population interactions. These models offer a fresh perspective, suggesting that what seems unpredictable is a natural outcome of complex adaptive systems moving along the edge of chaos.

Considering the practical implications of chaos in predator-prey interactions, parallels can be drawn with financial markets, where erratic stock price movements are influenced by underlying chaotic processes. Just as traders use sophisticated algorithms to forecast market trends, ecologists are developing predictive models to anticipate changes in animal populations. By incorporating chaotic variables, these models enhance prediction accuracy, allowing wildlife managers to create strategies that mitigate negative impacts on biodiversity. This highlights the value of chaos theory in guiding management decisions, promoting a proactive approach to ecological conservation.

To better understand this concept, imagine a scenario where a small increase in prey numbers temporarily boosts predator populations. While initially advantageous for predators, this increase eventually depletes the prey, leading to a subsequent decline in predator numbers. Over time, this cyclical pattern grows more intricate, with minor perturbations leading to unexpected outcomes. This thought experiment emphasizes the importance of considering chaos as a crucial component of ecological modeling, encouraging systematic exploration of how small changes can escalate throughout ecosystems.

Engaging with these ideas prompts reflection on the broader implications of chaos theory beyond ecology. In an interconnected world, recognizing how small changes can ripple through complex systems is essential. By leveraging this knowledge, individuals and communities can better navigate and influence the cascading effects present in various fields, from environmental management to economic planning. This approach not only fosters resilience but also empowers us to embrace the inherent unpredictability of our world as a driver for innovation and adaptation.

Population Synchronization and Collective Behavior

The phenomenon of population synchronization and collective behavior creates a captivating intersection of biology, mathematics, and sociology, revealing how individuals within a group can act in harmony without centralized guidance. This is not just a spectacle to behold but a demonstration of the fine-tuned balance within natural systems. At its essence, synchronization among populations is rooted in the interactions between individuals and their surroundings, resulting in emergent phenomena that exceed the sum of their parts. This is beautifully exemplified in the mesmerizing formations of starling murmurations or the synchronized flashes of fireflies. These occurrences are governed by simple interaction rules, yet they produce a complex choreography adaptable to shifting conditions and resilient against disruptions.

Recent research has shed light on the mechanisms that enable such synchronization, highlighting the significance of phase-coupling and feedback loops. For fireflies, synchronization in their flashing arises from a feedback process where each insect adjusts its timing based on the flashes of nearby peers. This decentralized control allows the group to achieve harmony without a leader. Similarly, fish schools and bird flocks exhibit alignment and cohesion through local interactions, with each individual responding to the movements of its neighbors. These self-organizing systems underscore the importance of local rules in shaping global structures, offering a blueprint for innovations in robotics and artificial intelligence, where autonomous agents must coordinate without central oversight.

The implications of population synchronization extend into human societies, where collective behavior can shape social norms, economic trends, and even political movements. Social synchronization is evident in the spread of ideas and behaviors, propelled by network effects and social feedback loops. For instance, the rapid flow of information across digital platforms can lead to synchronized public opinions, impacting elections and influencing cultural narratives. Understanding these dynamics equips us with tools to promote positive social change, leveraging synchronization to foster collaboration and collective action toward common goals. Harnessing these principles can lead to more effective strategies for mobilizing communities and addressing global challenges like climate change and public health crises.

From a theoretical standpoint, population synchronization challenges conventional ideas of predictability and control, prompting us to reconsider how order emerges from disorder. Chaos theory, with its emphasis on sensitivity to initial conditions, provides a valuable framework for exploring these dynamics. In predator-prey systems, synchronization can lead to cyclical population fluctuations, where the rise and fall of species numbers reflect a delicate balance of ecological forces. This interplay between disorder and order highlights the complexity of natural systems, raising questions about the predictability of such phenomena and the factors that influence their stability. Embracing this complexity, researchers are discovering new methods to model and predict the behaviors of synchronized populations, offering insights that could transform fields from ecology to economics.

For practitioners and researchers aiming to apply these insights, the key lies in identifying the simple rules that drive complex behaviors and leveraging these patterns to inform decision-making. Whether developing software for autonomous vehicles or designing interventions for community health, understanding population synchronization provides a robust framework for fostering coordination and enhancing system resilience. By studying the emergent properties of synchronized systems, we can devise strategies that align individual actions with collective objectives, ensuring that small, mindful actions contribute to meaningful, large-scale transformations. The challenge lies in translating these theoretical insights into practical applications, a pursuit that promises to unlock new potential for innovation and societal advancement.

Examining the central ideas of disorder and regularity, this chapter reveals the complex interaction between stability and chaos that exists in numerous systems. From the unpredictable twists of weather phenomena to the volatile swings of financial markets, and even the shifting behaviors of ecological populations, the balance between chaos and structure illustrates the deep-seated complexity and interdependence of our environment. These systems, which may appear random but are guided by fundamental principles, prompt us to reconsider our notions of control and anticipation.

By identifying patterns within disorder, we can understand how minor adjustments might lead to substantial transformations, highlighting the strength of intentional action. This investigation not only celebrates the beauty found in complexity but also equips us to apply these insights for informed decision-making. As we progress, the subsequent chapter will delve deeper into the significant ramifications of these concepts, encouraging readers to continue their exploration with a greater appreciation for the delicate equilibrium that defines our world.

Chapter Ten
Symmetry And Breaking Patterns

Observing the core of nature's most refined designs, one finds symmetry to be both a foundational guide and a delicate balance, inevitably prone to change. From the enchanting design of a snowflake to the complex weave of a spider's web, symmetry serves as a universal language, expressing harmony within complexity. Yet, it is in the moments of symmetry's disruption that new structures, ideas, and phenomena often arise. This intricate dance of balance and transformation is a timeless narrative, urging us to explore how these patterns ripple across the tapestry of existence.

Picture a universe where every particle aligns without flaw, each organism develops with exact precision, and cities expand with immaculate order. Such an idealized vision, though alluring, diverges from the reality we experience. In the microscopic world of particle physics, symmetry provides a framework to grasp the universe's fundamental forces, yet it is in the breaking of these symmetries that particles acquire mass and the universe as we know it forms. Similarly, in biological development, an embryo's symmetrical beginnings evolve into the intricate forms of life through calculated asymmetries. Urban landscapes begin with seemingly symmetrical designs, but as cities grow, these patterns adapt, reflecting the needs and influences of their inhabitants.

This chapter invites readers to delve into how the interplay between symmetry and asymmetry shapes the evolution of systems at every scale. By exploring examples from the subatomic to urban settings, we uncover the hidden forces driving change and innovation. These insights illuminate the principles of order and disorder, empowering us to identify and harness these dynamics within our environments. As we journey through this exploration, the universality of these patterns stands as a testament to the profound connections linking our smallest actions to the grandest transformations, reinforcing the book's overarching theme of small behaviors inspiring global change.

Particle Physics Symmetries

To truly appreciate the remarkable influence of balance in the field of particle physics, imagine a universe where unseen harmonies dictate the behavior of subatomic entities. At first, the cosmos may appear as a chaotic mix of matter and energy. Yet, beneath this superficial disorder lies a beautifully crafted design, governed by elegant proportions. These principles act as nature's blueprint, steering the interactions and transformations of particles. From maintaining equilibrium among forces to conserving energy, these patterns form the backbone of the universe. Understanding these designs allows scientists to unravel the mysteries of the cosmos, showing how the simplest harmonies lead to the intricate structures we witness.

This section invites readers on a journey to explore how fundamental balance shapes particle interactions, where each change adheres to specific conservation laws. We delve into the captivating domain of gauge symmetries, from which forces naturally arise to maintain equilibrium. The narrative then uncovers the subtle yet powerful phenomenon of spontaneous symmetry breaking, which leads to the Higgs mechanism and imparts mass to particles. As our exploration progresses, we venture beyond the Standard Model, glimpsing the enticing possibilities of supersymmetry and string theory, where new dimensions of balance emerge. Each topic reveals a layer of this intricate interplay, connecting micro-level interactions with the grand design of the universe, setting the stage for a deeper appreciation of balance's role in shaping our world.

Fundamental Role of Symmetry in Particle Interactions

Symmetry is a foundational concept in particle physics, serving as a guiding principle for understanding the interactions and behaviors of subatomic particles. It sheds light on the conservation laws governing these interactions, acting as a cornerstone of modern physics. Through symmetry, we can grasp how particles interact, combine, and disintegrate, revealing an underlying order in the apparent chaos of the subatomic realm. This profound influence not only helps classify particles but also predicts new ones, as seen in the prediction and eventual discovery of the Higgs boson. Once purely theoretical, the Higgs boson was brought into reality through the mathematical elegance of symmetry principles, which hinted at its existence long before technology could confirm it.

Exploring gauge symmetries has transformed our comprehension of fundamental forces, weaving them into a unified tapestry. These symmetries form the backbone of the Standard Model, governing electromagnetic, weak, and strong forces. They dictate the rules of particle interaction, ensuring certain properties remain constant. This adherence to symmetry is not merely a mathematical curiosity but a natural law governing how particles like quarks and leptons navigate the quantum field. The intricate equations of gauge symmetries unveil profound truths about the universe, such as charge conservation and the invariance of physical laws across different frames of reference.

Spontaneous symmetry breaking reveals the universe's dynamic nature. This phenomenon occurs when a system's symmetrical state becomes unstable, leading to a new, less symmetrical state. The Higgs mechanism illustrates this, explaining how particles acquire mass. In the early universe, symmetry breaking was pivotal, setting the stage for the formation of atoms, stars, and galaxies. This transition from symmetry to asymmetry resembles a phase change, where a subtle shift can lead to profound transformations. The universe's tendency for symmetry breaking hints at a deeper principle, suggesting potential advancements beyond current theoretical frameworks.

Beyond the Standard Model, theories like supersymmetry and string theory offer glimpses into unexplored territories. Supersymmetry proposes a symmetry between fermions and bosons, suggesting each particle has a heavier "superpartner." Although not yet experimentally confirmed, this concept could address some of the Standard Model's limitations, such as the hierarchy problem. String theory, alternatively, posits that particles are not point-like but rather tiny vibrating strings, with different frequencies corresponding to different particles. These advanced theories aim to unify all fundamental forces, potentially leading to a "theory of everything." As researchers delve into these ideas, they challenge conventional wisdom, inviting us to ponder the universe's profound mysteries.

Applying symmetry principles in practical scenarios can enrich our approach to designing resilient networks in technology or ecology, ensuring balanced interactions that prevent systemic failures. By understanding symmetry's role in various fields, we can devise innovative strategies, like optimizing resource distribution in urban planning or developing robust algorithms in computer science. Examining the underlying symmetry in these systems unlocks new possibilities for innovation and sustainability, fostering a mindset that appreciates the interconnectedness of seemingly disparate phenomena.

Exploring Gauge Symmetries and Conservation Laws

Gauge symmetries act as the unseen forces that knit the fundamental interactions of particles together, ensuring coherence in the universe at its most basic level. These symmetries, integral to space-time's fabric, uphold conservation laws that remain untouched through countless interactions. For instance, the conservation of electric charge naturally arises from the electromagnetic gauge symmetry, a principle reflected in phenomena ranging from atomic stability to electricity in everyday devices. This foundational concept influences not just electromagnetism but also the forces governing atomic nuclei, orchestrating the interactions of particles that define matter's essence.

Recent explorations into gauge symmetries have yielded groundbreaking insights in theoretical physics, particularly concerning the unification of forces. Symmetry isn't just a mathematical notion; it guides the development of grand unified theories aiming to reconcile electromagnetic, weak, and strong nuclear forces within a single framework. This pursuit of unity has sparked innovative approaches, like quantum field theories that place gauge symmetries at their core. These theories provide tantalizing glimpses into a potential underlying harmony in the universe's complexity, suggesting that simplicity might be the ultimate truth.

Beyond theoretical constructs, gauge symmetries hold practical significance in particle accelerators and high-energy physics experiments. The precision with which these symmetries predict particle interactions enables physicists to design experiments probing fundamental existential questions. The Large Hadron Collider, for instance, uses these principles to explore conditions similar to those following the Big Bang, offering empirical evidence that either supports or challenges existing theories. Its discovery of the Higgs boson, a particle predicted by symmetry principles, highlights the profound link between abstract symmetries and tangible reality, enhancing our grasp of mass and the forces shaping our universe.

Consider the intriguing implications of gauge symmetries within emerging frameworks like supersymmetry and string theory, which propose extensions beyond the Standard Model. Supersymmetry suggests that each known particle might have a heavier, yet undiscovered partner, proposing a symmetry between fermions and bosons that could solve pressing physics puzzles like dark matter's nature. Meanwhile, string theory posits particles as vibrating one-dimensional "strings," manifesting as different particles. These advanced theories rely on intricate symmetries to propose solutions to enduring riddles, including reconciling general relativity with quantum mechanics.

In the spirit of exploration and application, one might ponder how gauge symmetry principles could inspire innovations beyond physics. Imagine these concepts influencing engineering, where balance and conservation inherent in symmetry guide more efficient system designs, or computer science, where symmetry-based algorithms optimize data processing. By embracing the logic of symmetries, we cultivate a mindset that seeks to understand the universe's fundamental laws and applies these insights to foster advancements across diverse fields. Through the lens of gauge symmetries, we are reminded of the profound interconnectedness of all things and the potential for small, principled actions to catalyze transformative change.

Spontaneous Symmetry Breaking and Higgs Mechanism

Symmetry plays a pivotal role in particle physics, offering a fascinating dual nature: it both dictates the fundamental laws governing particles and, when disrupted, leads to intriguing phenomena like spontaneous symmetry breaking. This concept is crucial for understanding how elementary particles gain mass. In quantum physics, the Higgs mechanism illustrates this idea beautifully. Imagine an all-pervasive field throughout the universe that endows particles with mass as they move through it, similar to how a celebrity draws attention while walking through a crowd. This mechanism not only clarifies the origin of mass but also underscores the delicate balance between symmetry and its disruption, fundamentally shaping our universe.

The 2012 discovery of the Higgs boson at CERN confirmed this mechanism, supporting years of theoretical predictions. It provided direct evidence that particles obtain mass through interactions with the Higgs field. This discovery was transformative, enhancing our comprehension of particle physics and reinforcing the Standard Model, which explains electromagnetic, weak, and strong nuclear forces. Yet, this breakthrough also opens the door to further investigation, inviting physicists to explore how symmetry breaking impacts the universe's beginnings and development. The Higgs boson exemplifies how an abstract concept can have significant, concrete effects on our understanding of reality.

Beyond the Standard Model, spontaneous symmetry breaking paves the way for new theories, like supersymmetry. Supersymmetry suggests each particle has a superpartner, potentially addressing theoretical challenges such as the hierarchy problem. It proposes that the symmetry we perceive is just a glimpse of a deeper symmetry at higher energies. Although this framework remains under experimental evaluation, it could provide insights into dark matter and the early universe's conditions. The interaction between visible symmetry and its disruption compels physicists to reconsider fundamental principles governing particle interactions and the universe's composition.

Symmetry breaking's implications extend beyond theoretical physics, impacting areas like cosmology and materials science. In cosmology, it offers explanations for the early universe's phase transitions, possibly accounting for the matter-antimatter imbalance. In materials science, it informs the creation of new materials with specific properties, as seen in superconductors and ferromagnets. By understanding how breaking symmetry alters system properties, researchers can devise innovative solutions to complex challenges, from energy efficiency to quantum computing. This interdisciplinary influence highlights the universal relevance of symmetry principles and their disruptions.

As we delve deeper into particle physics, grasping spontaneous symmetry breaking remains essential. It challenges us to see how these principles manifest across various scales and contexts, guiding inquiries into the fundamental questions of existence. By embracing diverse approaches and methodologies, we can unravel the complex tapestry of symmetry and its breaking, gaining profound insights into the universe's workings. This journey not only advances scientific knowledge but also fosters a broader appreciation for the interconnectedness of all things, encouraging readers to explore how small, intentional disruptions can lead to transformative understanding and innovation.

Beyond the Standard Model: Supersymmetry and String Theory

In the fascinating field of particle physics, the exploration beyond the Standard Model invites us to consider the groundbreaking ideas of supersymmetry and string theory. These sophisticated frameworks aim to offer deeper insights into the universe's fundamental forces and propose elegant solutions to longstanding puzzles. Supersymmetry, often abbreviated as SUSY, suggests a relationship between bosons and fermions, indicating that each known particle might have an undiscovered partner. This symmetry could potentially address the hierarchy problem by stabilizing the Higgs boson's mass, preventing it from becoming excessively large. The pursuit of supersymmetry has led to numerous experimental efforts, with the Large Hadron Collider playing a pivotal role in the search for evidence of these theoretical partners.

String theory, an even bolder proposition, envisions particles not as point-like entities but as one-dimensional strings vibrating at unique frequencies. This conceptual shift offers a unified framework for gravity and quantum mechanics, potentially bridging the gap between general relativity and quantum field theory. Strings, through their dynamic oscillations, manifest as diverse particles, with their vibrational modes shaping the universe's intricate structure. The theory's allure lies in its potential to integrate all fundamental forces into a cohesive whole, extending its reach across multiple dimensions, some of which may be beyond our perceptual limits.

The interaction between supersymmetry and string theory creates a fertile ground for theoretical exploration. While SUSY may serve as a stepping stone, string theory expands the narrative, suggesting a multiverse of possibilities. The concept of extra dimensions, integral to string theory, invites profound reflection. Although these dimensions remain unseen, they could significantly impact particle interactions and cosmological phenomena. The quest for tangible evidence of these theories, such as supersymmetric particles or variations in gravitational behavior, continues to drive innovative experiments and technological advancements, pushing the boundaries of human knowledge.

Despite their appeal, both supersymmetry and string theory face significant challenges. The lack of empirical evidence for supersymmetric particles at current energy levels fuels ongoing debate and necessitates the refinement of theoretical models. Some physicists propose alternative frameworks, such as loop quantum gravity and emergent gravity, offering diverse perspectives on unifying nature's forces. This diversity of thought underscores the dynamic nature of the field, encouraging a multifaceted approach to unraveling the universe's deepest mysteries.

For those intrigued by the quest for understanding, these theories offer a glimpse into a cosmos both grand and intricate. Engaging with these concepts fosters a mindset of curiosity and critical thinking. Individuals can delve deeper into these ideas through accessible resources, online lectures, and collaborations with scientific communities. By cultivating a nuanced appreciation of these theoretical landscapes, one can meaningfully contribute to discussions that shape the future of scientific exploration.

Biological Development Patterns

Delving into the complexities of biological development reveals a captivating interplay of balance and diversity, essential to the fabric of life. At the core of this fascinating interaction is a masterful orchestration governed by nature's inherent principles. These developmental designs are not just visually pleasing—they are vital for the creation of structured forms and functionalities in living beings. From the swirling patterns of a sunflower to the mirrored wings of a butterfly, these designs are the product of ancient evolutionary processes honed over eons. This journey invites us to observe how such natural configurations arise, steered by genetic instructions and molecular signals that ensure life's seamless assembly.

Within this grace lies an equally intriguing phenomenon—the disruption of balance. These shifts are not mistakes but are crucial for diversity and adaptation. The journey from a single cell to a complex organism is marked by deliberate asymmetries that drive differentiation and specialization. The subtle changes in cellular congruity, the complex management of genetic expression, and the gradients of signaling molecules all play critical roles in shaping the vast array of life forms. These mechanisms craft not only individual organisms but also serve as models for understanding broader patterns, such as urban development and cosmic formations. By exploring cellular congruity, genetic management, signaling gradients, and evolutionary processes, we uncover the mysteries of how life's blueprints develop and transform, offering profound insights into the interconnectedness of all things.

Cellular Symmetry in Developmental Biology

In the intricate world of developmental biology, cellular symmetry plays a crucial role in shaping organisms. During embryonic development, symmetry is evident in processes like cleavage and differentiation, where cells follow precise patterns to form complex structures. This symmetry is not just for aesthetics; it is vital for correct and efficient development. Current studies reveal that molecular signals and environmental factors guide cellular symmetry, orchestrating the spatial arrangement of cells. This dynamic interaction between genetic data and external cues highlights how balance and harmony govern the architecture of life.

Recent advancements in imaging technologies have deepened our understanding of cellular symmetry. High-resolution microscopy and genetic labeling allow scientists to observe early cellular division with remarkable clarity. These tools have uncovered symmetry-breaking events—key moments when uniformity gives way to diversity, leading to specialized tissues and organs. Such discoveries underscore symmetry's importance in maintaining order while enabling deviations that fuel biological innovation. This duality, offering stability and change, invites researchers to explore the complex signals that drive these processes.

The study of cellular symmetry also intersects with genetic regulation, where specific genes are pivotal in establishing and maintaining symmetrical patterns. Genes like Sonic hedgehog and Nodal are essential for developing bilateral symmetry, guiding the left-right axis formation in vertebrates. Precise regulation of these genes ensures coordinated body structure development, preventing anomalies that could lead to disorders. Research into these genetic pathways not only enhances understanding of normal development but also opens doors for therapeutic interventions when symmetry is disrupted. As we explore the genetic basis of symmetry, breakthroughs in regenerative medicine and treatments for congenital disorders become increasingly promising.

Beyond traditional biology, cellular symmetry principles resonate in fields like bioengineering and synthetic biology. By imitating nature's symmetrical patterns, scientists are crafting innovative biomaterials and tissue engineering techniques. These applications leverage symmetry's inherent stability and efficiency to create structures that integrate seamlessly with biological systems. For instance, symmetrical scaffolds in tissue regeneration promote uniform cell growth and organization, improving healing outcomes. This interdisciplinary approach shows how understanding fundamental biological patterns can inspire technological advancements with significant implications for healthcare and beyond.

Looking to the future of cellular symmetry research, intriguing questions arise. How do cells interpret and respond to the myriad signals dictating symmetrical patterns? What potential exists for influencing developmental processes in a controlled manner? These inquiries fuel scientific curiosity and invite us to consider symmetry's broader implications in shaping life. By appreciating cellular symmetry's elegance and complexity, we gain insights into the natural world's governing principles, empowering us to apply this knowledge in ways that foster innovation and progress across various domains.

Genetic Regulation and Pattern Formation

Genetic regulation plays a crucial role in the development of biological patterns. At the core of this process are gene networks that function like a symphony, with each gene acting as a musician contributing to a harmonious outcome. Advances in systems biology have shed light on the precision of these genetic networks, which often regulate themselves through feedback mechanisms to maintain stability and adaptability. For example, the interactions between Hox genes, which determine an organism's body plan, showcase how genetic regulation can lead to complex, ordered patterns. These genes are not just on-off switches but dynamic components that respond to internal and external signals, demonstrating adaptability that allows organisms to flourish in varying environments.

In developmental biology, genetic regulation significantly influences pattern formation by controlling the timing and location of gene expression. This expression is not random but follows intricate rules honed over millennia. Take, for instance, the development of Drosophila melanogaster, the fruit fly, which serves as a model for understanding genetic patterning. The segmentation genes in Drosophila demonstrate a tiered system, where an initial gradient of maternal effect genes sets broad areas, which are then refined by gap genes and further delineated by pair-rule and segment polarity genes. This hierarchical regulation ensures precise structure formation, highlighting the power of genetic architecture in shaping biological forms.

Recent studies reveal that genetic regulation is not a fixed blueprint but a dynamic process shaped by epigenetic modifications and non-coding RNAs, adding complexity to pattern formation. Epigenetic changes can modify gene expression without altering the DNA sequence, enabling cells to respond flexibly to developmental signals. DNA methylation and histone modification, for instance, can activate or repress genes involved in cell differentiation and patterning. Meanwhile, non-coding RNAs, once considered non-functional, are now recognized as essential regulators that fine-tune gene expression and developmental patterns. These findings underscore the multifaceted nature of genetic regulation and its role in generating diversity in form and function.

The interaction between genetic regulation and environmental factors further complicates pattern formation. Environmental cues like temperature, light, and nutrition can influence gene expression patterns, leading to phenotypic plasticity. This adaptability is seen in phenomena like polyphenism, where a single genotype can produce multiple phenotypes based on environmental conditions. For example, the water flea Daphnia can develop protective spines or remain smooth-bodied depending on predator presence, illustrating how genetic regulation and environmental interaction shape developmental outcomes. This environmental integration highlights the evolutionary advantage of flexible genetic regulation in pattern formation.

Understanding genetic regulation of pattern formation opens doors to potential applications in fields like regenerative medicine, agriculture, and synthetic biology. Imagine engineering crops that adapt to climate change or developing therapies that guide tissue regeneration. By embracing the complexity of genetic networks, we can strive for a future where manipulating genetic regulation not only uncovers the secrets of life's patterns but also empowers us to address humanity's greatest challenges. This exploration encourages us to consider both known and uncharted territories awaiting discovery.

Morphogen Gradients and Spatial Organization

In the complex realm of developmental biology, morphogen gradients are fundamental in shaping spatial organization. These gradients, which are essentially concentration profiles of signaling molecules, direct cellular destinies, ensuring that differentiation occurs in an orderly and functional manner. The beauty of this system lies in its simplicity; a single gradient can convey positional cues to cells, helping them discern their location within a developing organism. Through this mechanism, morphogens like Sonic hedgehog and Bone morphogenetic proteins act as molecular architects, crafting the body plan of organisms from what initially appears as a chaotic cluster of cells.

Recent studies have shed light on the intricate ways cells decode these gradients. It is not solely the concentration of a morphogen that influences cellular behavior, but also the timing and interactions with other signaling pathways. This interplay can result in complex differentiation patterns, akin to a symphony where every note plays a role in the overall masterpiece. Research in model systems like the fruit fly Drosophila has revealed how minor variations in morphogen concentrations can lead to significant differences in pattern formation, highlighting the sensitivity and precision of this biological process.

The concept of morphogen gradients also finds intriguing applications beyond traditional biology, offering parallels in fields such as urban planning and network theory. Just as cells organize themselves based on morphogen cues, urban growth can be shaped by factors creating "gradients" of economic opportunity or social connectivity. Recognizing these parallels can provide innovative strategies for designing sustainable cities that are both efficient and harmonious. By applying principles of morphogenetic patterning, urban planners can forecast growth patterns and address challenges associated with rapid urbanization.

Advances in biotechnology now allow scientists to manipulate morphogen gradients with remarkable precision. Techniques like optogenetics and CRISPR-based gene editing enable the fine-tuning of morphogen expression, unlocking new possibilities in regenerative medicine and tissue engineering. These developments not only expand the potential for repairing damaged tissues but also pave the way for creating synthetic organs with the same precision as those naturally formed. The implications of these technologies are profound, promising to revolutionize our approach to complex medical challenges.

As we contemplate the future of morphogen research, it is essential to consider the ethical and philosophical questions it raises. With the ability to engineer life so precisely, what responsibilities do we have in shaping the organisms of tomorrow? How might these capabilities change our understanding of life's complexity and diversity? By pondering these questions, readers are encouraged to think critically about the intersection of science, society, and ethics, and explore how the principles of morphogen gradients might inform their own efforts to foster meaningful change.

Evolutionary Mechanisms in Pattern Diversification

The vast array of designs in biological development highlights the complex interplay of evolutionary processes. Through this dynamic framework, understanding how pattern diversification through evolution occurs sheds light on the adaptability and success of organisms. Core evolutionary processes, including natural selection and genetic drift, are instrumental in crafting the diverse arrangements seen in nature. These mechanisms act on genetic variations within populations, favoring traits that improve survival and reproduction. Over time, this results in the emergence of unique designs tailored to specific ecological niches. The interaction between genetic diversity and environmental pressures exemplifies life's enduring innovation to maintain its presence in a shifting world.

In the realm of developmental biology, the variation of designs is also molded by epigenetic factors and gene regulation networks. Epigenetics enables organisms to quickly adapt to environmental shifts without changing their DNA sequence, offering a versatile mechanism for design adaptation. For example, the ability of insects to produce varying phenotypes from the same genetic code—a phenomenon termed phenotypic plasticity—illustrates how organisms can adjust their developmental processes to suit different conditions. This potential for phenotypic adjustment highlights the evolutionary benefit of maintaining a repertoire of possible designs, ready to be employed when needed.

The study of morphogen gradients further clarifies how evolutionary processes drive design diversification. Morphogens, signaling molecules diffusing through developing tissues, create concentration gradients that determine cell fate and tissue structuring. Variations in these gradients can lead to a wide range of morphological outcomes, propelling the evolution of new forms. Investigations into the genetic and molecular basis of these gradients reveal how slight changes in morphogen distribution or receptor sensitivity can result in significant morphological innovations. These discoveries underscore the role of evolutionary experimentation in producing the extensive variety of life forms observed globally.

Cutting-edge research in evolutionary developmental biology (evo-devo) reveals the interconnected nature of evolutionary and developmental processes in design diversification. Scientists are discovering how alterations in developmental pathways can lead to evolutionary innovations, providing insight into the mechanisms driving speciation and adaptation. For instance, examining limb development in vertebrates has shown how changes in the timing and expression of crucial developmental genes can lead to the evolution of new limb structures. These findings demonstrate the power of evolutionary processes to create novel designs by repurposing existing developmental pathways.

Applying these insights practically, evolutionary principles can inform fields such as biomimicry and synthetic biology. By understanding the mechanisms behind design diversification, researchers and innovators can craft systems that emulate nature's adaptability and resilience. Designing materials or structures that respond dynamically to environmental stimuli could revolutionize industries from architecture to healthcare. Fostering a mindset that embraces evolutionary thinking can inspire innovative solutions to complex challenges, emphasizing the significant impact that understanding design diversification can have on both science and society.

Urban Growth Models

Picture waking up to a city that seems to have reshaped itself overnight, expanding in unexpected ways with new neighborhoods emerging like a burst of wildflowers. This seemingly haphazard growth is, in fact, guided by a complex web of mathematical principles. Urban areas, much like ecosystems, grow in patterns that are both orderly and chaotic, reflecting a hidden symmetry akin to the forces shaping the natural world. Just as living organisms evolve, cities transform through influences like historical development, spatial dynamics, and the continuous ebb and flow of human activity. Delving into urban growth models reveals the unseen order that guides cities from their early stages to vast metropolises, echoing the broader principles seen throughout the cosmos.

Urban expansion mirrors the mathematical framework that underpins our universe. From ancient towns to today's sprawling cities, these environments are a fusion of historical evolution and precise mathematical design. Invisible yet powerful feedback mechanisms drive urban development, affecting infrastructure and population spread. Fractal geometry, meanwhile, offers a glimpse into future possibilities, aiding in sustainable urban planning by predicting growth patterns. As we explore these fascinating concepts, we see how cities mimic natural and cosmic systems, all shaped by the universal language of patterns, connecting the tiny details to the grand design in a continuous cycle of growth and transformation.

Historical Evolution of Urban Spatial Patterns

The evolution of urban spatial patterns highlights the dynamic interaction between human innovation and environmental limitations. Early cities, strategically located near rivers or fertile lands, often exhibited a radial design focused around a central marketplace or religious hub. This setup was not only a reflection of societal organization but also a practical adaptation to local geography and resources. With the advent of the Industrial Revolution, a significant shift occurred in urban design, leading to the widespread adoption of grid systems. This transformation was motivated by the need for efficient transportation networks and to support the burgeoning industries. Cities like New York exemplify this grid approach, enabling predictable expansion and accommodating rapid population growth, driven by economic prospects.

Exploring the mathematical basis of urban growth reveals fractal geometry as a crucial framework. Fractals, with their self-repeating patterns at various scales, closely resemble the organic expansion of cities. The seemingly chaotic yet predictable extension of suburbs, known as urban sprawl, can be analyzed using models that simulate growth patterns. These models offer insights into how cities expand in fractal dimensions, balancing population density with accessibility. By comprehending these fractal dimensions, urban planners can better forecast congestion areas and design more sustainable urban environments that optimize space use and infrastructure efficiency.

Feedback loops significantly influence urban growth, acting as both accelerators and regulators. Positive feedback loops, like the clustering of businesses in high-density areas, can spur urbanization and economic growth. In contrast, negative feedback loops, such as resource strain in overcrowded districts, can lead to the dispersal of residents and businesses. Recognizing these dynamics allows policymakers to craft strategies that amplify positive trends and counteract negative impacts. For example, zoning laws promoting mixed-use development can foster self-reinforcing cycles of growth and sustainability, seamlessly integrating residential, commercial, and recreational spaces.

The advancement of predictive modeling, enhanced by fractal analysis, provides groundbreaking insights into urbanization's future. These models, refined by technological progress in computational power and data analytics, enable highly accurate simulations of urban growth scenarios. By accounting for factors like population density, resource availability, and environmental impact, urban planners can anticipate and adjust to potential challenges. This forward-thinking approach not only facilitates strategic planning but also builds resilience and adaptability against unforeseen events, ensuring cities remain vibrant and sustainable centers of human activity.

As we delve into the historical progression and mathematical principles governing urban spatial patterns, it is essential to ponder questions that challenge existing paradigms. How can cities leverage fractal geometry to revolutionize urban design? What strategies can balance the inherent feedback loops in urban systems? Such questions invite a critical examination of the intersection between mathematics and urban planning, promoting a comprehensive approach to shaping the cities of the future. By incorporating diverse perspectives and advanced insights, we can cultivate urban environments that embody the complexity and elegance of the patterns that guide their growth.

Mathematical Foundations of Urban Growth Dynamics

In the ever-evolving landscape of urban development, the mathematical principles form the core of understanding how cities grow and transform. A fascinating perspective is offered by complex systems theory, which perceives urban environments as intricate webs of interactions and dependencies. These systems are marked by emergent behavior, where minor changes can lead to substantial effects. The concept of self-organization, often seen in nature, is mirrored in urban development as cities naturally adapt and reorganize in response to population increases and economic forces. By studying these self-organized structures, researchers can pinpoint the main factors driving urban expansion and forecast future developments.

Fractal geometry, which explains self-repeating patterns, provides valuable insights into urban growth dynamics. Cities often display fractal traits, where smaller configurations replicate at larger scales. This is evident in the way residential areas mimic the layout of larger districts or how road networks resemble the veins of a leaf. This fractal quality aids in effective urban planning and offers a framework for anticipating changes in land use and infrastructure demands. Through the use of fractal models, urban planners can predict city expansion and plan infrastructure development to support sustainable growth.

The impact of mathematical models extends to exploring urban growth through cellular automata, a computational technique that models how simple rules applied to individual units result in complex patterns. By simulating the gradual spread of urban areas, cellular automata allow for the exploration of various growth scenarios, such as the effects of zoning regulations or transportation systems on future development. This method is invaluable for planners and policymakers, offering a visual and analytical tool to test theories and assess the outcomes of different urban policies.

Recent advancements in machine learning have further enhanced the mathematical underpinnings of urban growth dynamics. Algorithms capable of processing vast datasets now uncover patterns and trends that were previously undetectable. For instance, by examining satellite imagery and socioeconomic data, machine learning models can identify urban growth hotspots, enabling timely intervention and planning. This ability to harness data-driven insights allows cities to proactively address challenges linked to rapid urbanization, such as resource management and environmental sustainability.

As cities continue to evolve and adapt, the mathematical exploration of urban growth dynamics remains a field ripe for innovation. By merging traditional mathematical models with modern technology, there is potential to revolutionize urban planning practices. This synergy of mathematics and technology not only enhances our understanding of cities but also empowers us to create urban environments that are resilient, efficient, and in harmony with natural ecosystems. As readers reflect on these insights, they are encouraged to consider how these mathematical principles could be applied to their own urban contexts, fostering a deeper connection with the cities they inhabit.

The Role of Feedback Loops in Urban Expansion

Urban expansion unfolds as a captivating narrative driven by the complex dynamics of feedback mechanisms. These cycles are fundamental in shaping the vast landscapes of contemporary cities, influencing growth patterns that mirror and intensify human activity. As urban areas develop, they transform into dynamic entities where every decision, from zoning regulations to infrastructure projects, resonates throughout the city, impacting future progress. This self-reinforcing characteristic of urban expansion resembles a perpetual cycle where initial changes in land use or policy trigger a series of adjustments. Advanced computational tools now empower urban planners to model these feedback processes, providing insights into how small changes can lead to significant shifts over time.

Central to these feedback mechanisms is the intricate balance between supply and demand. As cities grow, the need for resources such as housing, transportation, and public services rises, spurring further development to fulfill these demands. This growth, in turn, attracts more people and businesses, perpetuating the cycle. These feedback loops are not just theoretical concepts; they are observable phenomena, as seen in historical city development patterns. For example, the introduction of a new transit line can increase nearby property values and stimulate additional commercial and residential projects, demonstrating how infrastructure investments contribute to the growth cycle.

Recent studies highlight technology's role in influencing these feedback loops. Smart city projects, integrating data analytics and Internet of Things (IoT) devices, provide unparalleled control over urban growth patterns. By tracking real-time data on traffic, energy consumption, and environmental factors, city planners can make informed decisions to optimize urban development. This technological feedback boosts efficiency and promotes sustainability, ensuring cities grow in harmony with their environment. Additionally, digital twins—virtual models of physical urban areas—allow planners to test scenarios, enabling them to assess the impact of their decisions before real-world implementation.

Nevertheless, the positive aspects of feedback loops in urban growth come with challenges. Uncontrolled expansion can lead to urban sprawl, environmental harm, and socio-economic inequalities. To address these issues, innovative approaches advocate for a holistic perspective on urban planning, incorporating principles of equity, resilience, and environmental stewardship into the feedback loops that drive city expansion. By fostering inclusive growth, cities can become more livable, fair, and sustainable, ensuring that urbanization benefits all residents.

Considering the future of cities also encourages critical reflection on the potential for feedback loops to reshape urban environments. Imagine if these mechanisms could be harnessed to create self-regulating cities that dynamically adapt to changing conditions. Visualize a city where data-driven insights continually refine urban design, optimizing resource allocation and minimizing waste. As we move into the era of smart cities, the challenge is to balance technological innovation with human-centered values. By understanding and utilizing the power of feedback loops, urban planners and decision-makers can craft cities that not only prosper now but also anticipate and adapt to future challenges.

Predictive Modeling of Urbanization Using Fractal Geometry

Predictive modeling of urbanization through the lens of fractal geometry provides a revolutionary way to understand the intricate sprawl of contemporary cities. At its essence, fractal geometry uncovers self-repeating patterns that manifest at various scales, from the winding paths of historic towns to the towering structures of urban metropolises. Viewing cities as fractal constructs, where designs recur at different levels, enables us to forecast urban growth and change. This method equips urban planners and researchers with the foresight to predict expansion trends and develop strategies fostering sustainable growth. The fractal essence of urban environments highlights the interconnectivity of urban elements, offering profound insights into their functioning.

Integrating fractal geometry into urban planning requires acknowledging the repetitive motifs that typify city development. For example, the arrangement of facilities such as parks, educational institutions, and public transit often mirrors a fractal design, ensuring accessibility across different scales. Recognizing this allows planners to efficiently allocate resources, meeting the demands of bustling city centers and sprawling suburbs alike. By harnessing fractal geometry in predictive models, cities can strike a balance between expansion and livability, enhancing the urban experience while building resilience against demographic changes and environmental pressures.

Recent studies have unveiled the potential of fractal-based models to accurately simulate urban growth. These models have shown proficiency in predicting the spatial distribution of population density, infrastructure evolution, and even economic activities. By capturing the complex nature of urban systems, fractal models offer a deeper understanding of urban behavior than traditional linear methods. This empowers policymakers to devise strategies that anticipate growth patterns, alleviate congestion, and conserve green spaces. Accurately predicting urbanization trends opens new pathways for developing cities that are both efficient and sustainable.

A pivotal aspect of employing fractal geometry in urban research is its ability to expose the hidden structure of cities. By examining the fractal dimensions of urban forms, researchers can identify the underlying order within the seeming chaos. This perspective encourages creative reimagining of urban spaces, leading to innovative designs that mirror the organic growth patterns seen in nature. By adopting fractal principles, architects and urban designers can craft environments that seamlessly integrate with their natural surroundings, enhancing ecological balance and visual appeal. This approach challenges traditional urban development concepts, providing a visionary framework for future cities.

As urbanization accelerates, the significance of predictive modeling using fractal geometry becomes increasingly apparent. This approach equips stakeholders with essential tools for navigating the complexities of modern urban life, fostering adaptable, equitable, and vibrant spaces. Embracing the fractal characteristics of cities unlocks new opportunities for sustainable growth, ensuring urbanization acts as a catalyst for positive transformation. Through this perspective, small-scale interventions can ripple throughout the urban landscape, sparking significant changes that improve the quality of life for all residents. This paradigm shift invites us to view cities not just as clusters of buildings but as dynamic, thriving ecosystems full of potential.

This chapter explores the dynamic interplay between symmetry and the disruption of patterns, illustrating their fundamental role in shaping the universe—from the tiniest particles to vast urban landscapes. In particle physics, symmetrical laws elegantly govern matter, yet it's the disruption of these symmetries that births new particles and forces, highlighting the ever-evolving nature of the cosmos. In biological realms, symmetry orchestrates the harmonious growth of organisms, while slight asymmetries drive diversity and adaptation, demonstrating life's remarkable adaptability. Urban development models further exemplify how balance and its disruption influence city evolution, harmonizing uniformity with innovation to cultivate vibrant communities. These insights reveal symmetry's dual role as a stabilizing and transformative influence across various scales and contexts. As we contemplate the designs that shape our world, the challenge lies in discerning when to embrace balance and when to welcome change, empowering us to thoughtfully apply these principles in our endeavors. This understanding equips us to navigate the complexities of change, setting the stage for further exploration into the interconnected tapestry of universal designs.

Chapter Eleven
Energy Flow Systems

In the midst of a lively city, where each street pulses with its own unique rhythm, an unseen symphony unfolds, orchestrating a seamless flow of vitality. Picture the power lines resonating above, synchronized traffic lights guiding the flow below, and the vibrant exchange of ideas and emotions. This intricate dance of energy mirrors the fundamental forces that shape our universe, illustrating how power moves, transforms, and sustains everything from thriving urban centers to distant galaxies. Just as a city's lifeblood depends on the harmonious distribution of resources, the cosmos flourishes on a grand scale, linked by invisible strands of force.

Energy, whether coursing through the veins of living beings, fueling global economies, or igniting distant stars, acts as a universal currency underpinning all systems. It is the unseen architect of life's processes, shaping the reality we perceive. In this chapter, we explore the fascinating world of cellular metabolism, where life's alchemy converts nutrients into the vital force animating organisms. We then delve into the intricate networks of economic resource distribution, revealing parallels between biological systems and the markets that drive our world. Our journey concludes with a gaze towards the heavens, examining how celestial mechanics drive the vast universe.

These narratives of energy dynamics unveil a profound unity in diversity—how principles governing a single cell's metabolism echo through economic systems and celestial bodies. By uncovering these connections, we gain insight into the complex fabric of existence, where even the smallest actions ripple outwards, reshaping the world in subtle and profound ways. Understanding energy dynamics empowers us to harness these forces, fostering a future where mindful actions can ignite transformative change globally.

Cellular Metabolism

Delving into the dynamic world of cellular metabolism, we uncover the precise and elegant interplay of energy that fuels life's fundamental processes. Imagine a city brimming with activity, its avenues bustling as people and vehicles weave through a labyrinth of thoroughfares. Similarly, cells function as miniature cities, where enzymes act as diligent organizers, directing metabolic routes with remarkable skill. These pathways, the lifelines of the cell, facilitate a continuous flow of vitality, ensuring every cellular activity is powered and each minute component contributes to life's intricate fabric. This microscopic harmony not only sustains individual cells but also reflects the universal theme of energy movement, resonating across all levels of existence, from a single organism to the cosmos itself.

As we explore this complex landscape, the mechanisms of cellular respiration reveal the sophistication with which cells extract energy, transforming the latent potential in nutrients into a driving force that propels life. Here, the metabolic rate serves as a skilled conductor, harmonizing efficiency and speed to meet the ever-changing needs of the environment. This adaptability echoes the resilience found in ecosystems and economies, which shift strategies in response to challenges. With these insights, we uncover the deep interconnections of energy dynamics, where the small mirrors the vast, and the patterns of life's smallest units echo across the universe. Through this lens, we gain a richer understanding of how these cellular processes not only sustain life forms but also weave together the intricate web of energy flow that unites all scales of existence.

The Role of Enzymes in Metabolic Pathways

Enzymes, the biological catalysts that drive cellular metabolism, are essential for the efficient and precise execution of biochemical reactions. These protein molecules reduce the activation energy needed for reactions, thereby speeding up metabolic processes vital for life. Enzymes play a crucial role in metabolic pathways, akin to conductors orchestrating a symphony, ensuring each reaction occurs at the right moment and location. Advances in enzyme engineering have uncovered new potential, suggesting innovative uses, such as creating more efficient biofuels or novel therapeutic interventions.

The specificity of enzymes is a testament to their molecular design, honed over millions of years of evolution. Each enzyme is uniquely suited to its substrate, facilitating transformations that support cellular respiration, photosynthesis, and other essential metabolic activities. For example, hexokinase initiates glycolysis by phosphorylating glucose, a critical step in extracting energy from carbohydrates. Understanding this specificity not only deepens our appreciation of biological systems but also offers opportunities for biotechnological applications, such as designing enzyme inhibitors to combat diseases linked to metabolic disruption.

Enzymes demonstrate an impressive ability to adapt to varying cellular and environmental conditions, thereby maintaining homeostasis. This adaptability is evident in processes like enzyme induction, where specific substrates trigger increased enzyme production. Such mechanisms allow organisms to optimize resource usage and energy expenditure. For instance, the inducible enzyme lactase enables some mammals to digest lactose beyond weaning, a trait evolved in response to dietary needs. This adaptability highlights the dynamic nature of enzymes and underscores their evolutionary importance in aiding survival across diverse ecological settings.

The growing field of systems biology offers a comprehensive view of enzymes within complex metabolic networks. By merging computational models with experimental data, researchers can simulate how changes in enzyme activity affect overall cellular function. This systems-based approach has led to breakthroughs in understanding metabolic disorders and identifying new therapeutic targets. For example, insights into the enzyme dynamics of cancer cells have led to targeted therapies that disrupt abnormal metabolic pathways, offering promising avenues in cancer treatment. This interdisciplinary research exemplifies the transformative potential of merging traditional biology with computational sciences.

Reflecting on enzyme function prompts us to consider how this knowledge might address broader societal challenges. Enzyme research holds significant implications for global issues like food security and sustainable energy. Enzymes could be pivotal in developing crops with improved nutrient profiles or engineering microorganisms capable of breaking down pollutants. As we continue to unravel the mysteries of these molecular catalysts, the potential for innovative solutions to pressing challenges becomes increasingly viable. Through the study of enzymes, we are reminded of the profound impact that understanding and manipulating the microscopic world can have on the larger world, encouraging us to envision a future where biological innovation drives positive societal change.

Energy Harvesting Through Cellular Respiration

Cellular respiration is fundamental for energy production in cells, involving a complex series of biochemical reactions that transform nutrients into energy. This process mainly unfolds within the mitochondria, often dubbed the cell's powerhouses, and comprises several stages: glycolysis, the citric acid cycle, and oxidative phosphorylation. Each phase methodically extracts energy from glucose and other substrates, converting it efficiently into adenosine triphosphate

(ATP), the primary energy carrier in cells. This transformation is not merely a biological process but a crucial phenomenon underpinning the vitality of living organisms. Recent molecular biology advances have illuminated the dynamic regulation of these pathways, showing how cells optimize energy production to meet various demands, such as during high activity or environmental stress.

A key feature of cellular respiration is its impressive adaptability. Cells can switch between aerobic and anaerobic respiration based on oxygen availability, ensuring survival and energy production in changing environments. In oxygen-rich conditions, full oxidation of glucose occurs, maximizing ATP production. Conversely, when oxygen is limited, cells turn to anaerobic pathways like fermentation, which, despite being less efficient in ATP yield, are essential for survival. This adaptability highlights the evolutionary brilliance of cellular systems, enabling them to thrive in diverse habitats. Researchers are exploring this adaptability further, investigating how metabolic flexibility can be harnessed to develop therapies for diseases related to metabolic dysregulation, such as cancer and diabetes.

The efficiency of cellular respiration is closely tied to coenzymes like NAD+ and FAD, which serve as electron carriers in the electron transport chain. These molecules transport electrons from nutrient oxidation to the mitochondria's inner membrane, where a series of redox reactions creates a proton gradient. This gradient drives ATP synthesis through chemiosmosis, exemplifying nature's elegant use of electrochemical gradients for energy conservation. Recent research has focused on the structural intricacies of these coenzymes and their interactions, identifying potential targets for pharmaceutical intervention in metabolic disorders. Such studies not only enhance our understanding of cellular energetics but also open new pathways for therapeutic advancements.

Beyond the cellular level, the principles of energy conversion in respiration have significant implications for broader scientific fields. Concepts of energy transformation and efficiency resonate across disciplines, from ecology, where they inform ecosystem productivity models, to technology, where they inspire sustainable energy system designs. For instance, engineering approaches that mimic biological processes draw inspiration from the efficiency of cellular respiration to create more sustainable energy solutions. As we face increasing energy demands and environmental challenges, insights from cellular respiration provide a blueprint for developing adaptive, resilient solutions that align with nature's strategies.

Considering the complexity of cellular respiration, one might ask: How can these biological insights be harnessed to tackle global challenges? There is potential in bioengineering to optimize metabolic pathways for enhanced biofuel production or exploring new metabolic routes to mitigate climate change effects. These questions invite reflection on the intersections of biology, technology, and sustainability, encouraging a shift from passive understanding to active participation in using nature's principles for innovative change. Through thoughtful exploration and application, the foundational knowledge of cellular respiration can inspire transformative solutions extending beyond the cellular realm.

The Impact of Metabolic Rate on Cellular Efficiency

Metabolic rate, a fundamental aspect of cellular activity, governs the pace at which cells transform nutrients into energy, affecting growth and survival. Central to this process is the equilibrium between energy consumption and acquisition, a complex interaction of biochemical pathways that underpin life. A high metabolic rate can be beneficial, enabling cells to swiftly meet energy demands, but it requires increased resource intake. On the other hand, a lower rate might conserve energy but hinder rapid adaptation to environmental shifts. This delicate balance is seen in the diverse strategies of organisms, from the rapid growth of bacteria to the energy conservation in hibernating mammals.

Recent research has uncovered how cells fine-tune their metabolic rates for optimal efficiency. One mechanism involves regulating mitochondrial activity, where cells alter the number of active mitochondria based on energy needs. This dynamic regulation ensures energy production matches consumption, reducing waste and boosting productivity. Cutting-edge studies reveal that cells use genetic and epigenetic modifications to adjust mitochondrial function, opening new paths for understanding diseases like diabetes and cancer linked to metabolic issues.

Cellular metabolism's efficiency involves not just energy production but also its judicious use. Cells have developed sophisticated methods to prioritize essential pathways while minimizing non-essential ones. This selective resource allocation resembles an economic model where supply and demand influence activity levels. External pressures, such as nutrient scarcity or oxidative stress, can disrupt this balance, prompting adaptations that prioritize survival over efficiency. This dynamic interplay showcases cellular systems' resilience and adaptability to changing environments.

The impact of metabolic rate extends beyond individual cells, influencing larger biological systems. For example, the collective metabolic activity within an ecosystem affects resource distribution and ecological dynamics. At the cellular level, metabolic efficiency can drive evolution, favoring traits that optimize or reduce energy use. Understanding these connections illuminates natural selection and ecological balance, highlighting how microscopic metabolic strategies can shape macroscopic phenomena.

As research advances, practical applications of metabolic rate studies grow. By manipulating metabolic pathways, scientists are devising new methods to boost agricultural yields, develop biofuels, and treat metabolic disorders. These innovations underscore the transformative potential of understanding cellular metabolism, offering solutions to global challenges. Readers are encouraged to consider how principles of metabolic efficiency might inform their pursuits, whether in science, environmental care, or personal health. Exploring cellular energy landscapes not only reveals life's intricacies but also inspires steps toward a sustainable future.

Adaptive Metabolic Strategies in Response to Environmental Changes

Cells exhibit extraordinary versatility in managing energy through their intricate microscopic structures. This adaptability is crucial for survival, allowing them to recalibrate metabolic pathways in response to shifts in nutrient availability, temperature changes, and other environmental cues. A notable mechanism is metabolic plasticity, enabling cells to alternate between energy sources, such as glucose and fatty acids, based on their immediate surroundings. This flexibility not only preserves cellular functions but also enhances energy efficiency, similar to how an organism adjusts its diet according to available resources.

Recent breakthroughs in molecular biology have illuminated the regulatory networks behind these adaptive responses. Central to these processes are transcription factors and signaling molecules, which act as gatekeepers of metabolism. They detect environmental changes and adjust gene expression to modify metabolic pathways. For example, under low-oxygen conditions, hypoxia-inducible factors (HIFs) activate glycolysis pathways, allowing cells to maintain energy production when oxidative phosphorylation is impaired. This shift, though energetically demanding, underscores the cell's prioritization of survival over efficiency in challenging environments.

Beyond individual cellular responses, the interactions between cells within tissues or organ systems further demonstrate adaptive metabolic strategies. Cancer cells, for instance, frequently exploit altered metabolic pathways to support rapid growth and survival in harsh conditions like limited oxygen and nutrient supply. Known as the Warburg effect, this involves a preference for glycolysis over oxidative phosphorylation, even when oxygen is abundant. By understanding these pathways, researchers are developing targeted therapies to disrupt the metabolic flexibility of cancer cells, aiming to inhibit tumor growth.

The implications of adaptive metabolism extend across various biological and ecological systems. In ecosystems, organisms adjust their metabolism to thrive in diverse environments, from nutrient-scarce deserts to nutrient-rich rainforests. These adaptations are also reflected in human-designed systems, like urban planning and agriculture, where resource distribution and energy use must be optimized to meet changing conditions. Drawing parallels between cellular and systemic metabolic strategies offers insights into creating resilient and sustainable infrastructural solutions.

Studying adaptive metabolic strategies invites us to explore the profound interconnectedness of biological entities and their environments. It poses intriguing questions about how cells, as fundamental units of life, can inform broader ecological and societal systems. What can we learn from cellular adaptability to enhance our responses to climate change or resource scarcity? By applying these principles, we have the potential to foster innovation and resilience in the face of global challenges, ensuring not just survival but also thriving in a dynamic world.

Economic Resource Distribution

Economic systems resemble dynamic ecosystems, intricately woven with patterns that govern the flow of resources and shape market behaviors. These patterns are far from abstract; they dictate how resources are distributed, influence economic trends, and affect the overall well-being and stability of societies. Picture a vast network where each connection symbolizes a transaction, investment, or trade agreement, continuously evolving. Within this network, feedback mechanisms are crucial, establishing equilibrium and maintaining the delicate balance between supply and demand. Often unseen, these mechanisms can amplify or diminish economic activities, creating waves that lead to growth or decline. The movement of resources in economics is about balancing stability with adaptability.

As we delve into how resources are distributed economically, the forces of exponential growth and network effects emerge as significant players. While exponential growth spurs innovation and expansion, it challenges sustainable resource management, requiring creative strategies to mitigate its impact. Network effects add complexity by linking global trade systems, enhancing efficiency but also bringing challenges of dependency and resilience. Amidst this complexity, market dynamics reveal their unpredictable nature, where chaos theory sheds light on what seems random yet follows certain patterns. By understanding these forces, we can better predict economic changes and build systems that are robust and adaptable, paving the way for a more sustainable global economy.

The Role of Feedback Loops in Economic Equilibria

Feedback loops are essential in sustaining economic balance, serving as forces that guide markets toward stability. These loops involve processes where a system's output is reintroduced as input, affecting future outcomes. In economics, they appear in various ways, such as price changes due to shifts in supply and demand or interest rate adjustments prompted by inflation. This dynamic ensures markets can endure shocks and return to balance. Economists and strategists increasingly appreciate the complexity of feedback loops, aiming to leverage these mechanisms for economic stability.

A compelling example of feedback loops is seen in how central banks use monetary policy to manage inflation. When inflation increases, central banks may raise interest rates, typically reducing spending and investment, thus easing inflationary pressures. This illustrates a negative feedback loop, where the system corrects deviations to maintain balance. The adaptability of such systems is noteworthy; they can be finely adjusted using real-time data, allowing nuanced responses to complex economic challenges. This adaptability is vital in today's rapidly changing global economy.

Advancements in data analytics and machine learning have deepened our understanding of these feedback mechanisms. With vast datasets, economists and researchers can now model and predict economic behaviors with remarkable accuracy. This foresight allows for the early detection of potential imbalances, enabling proactive measures to counteract negative effects. For instance, algorithmic trading systems utilize feedback loops to evaluate market conditions and execute trades swiftly, highlighting how technology can apply these principles innovatively. Integrating artificial intelligence in economic systems hints at a future where feedback loops are even more refined, suggesting a more stable economic environment.

Nevertheless, not all feedback loops are advantageous. Sometimes, they can lead to negative outcomes, such as market bubbles or financial crises. Positive feedback loops, where outputs amplify inputs, can create runaway effects, resulting in volatility and systemic risk. The 2008 financial crisis starkly illustrates how unchecked positive feedback in housing markets and financial derivatives can lead to severe economic downturns. This dual nature emphasizes the need for vigilance and regulatory frameworks to identify and mitigate potential threats before they escalate.

To effectively harness feedback loops, decision-makers must adopt a comprehensive view, considering both micro and macroeconomic factors. Encouraging transparency and fostering communication among global economic entities can help synchronize actions to achieve desired outcomes. Understanding these principles empowers more informed decision-making, whether in personal finance, corporate strategy, or policy development. By cultivating an awareness of feedback loops and their broad implications, stakeholders can contribute to building a more resilient and equitable economic future.

Exponential Growth and Its Impact on Resource Allocation

Exponential growth, a phenomenon marked by rapid expansion, significantly influences how resources are distributed within economic systems. As industries, populations, or financial markets expand at accelerating rates, the demand for resources increases dramatically, challenging traditional distribution methods. This swift growth often results in resource scarcity, prompting the need for innovative allocation strategies. Recent research highlights the transformative role of digital currencies and blockchain technology in resource distribution, offering decentralized and efficient transaction models. These technological advancements promote fair access to resources, addressing the imbalances commonly linked to exponential growth.

Managing resources amid exponential growth extends beyond logistics; it requires understanding the fundamental drivers of such growth. Technological innovation, demographic changes, and policy shifts can all act as catalysts, shaping resource distribution. For example, the surge in renewable energy technologies demonstrates how exponential growth in one sector can influence resource allocation in others. As solar and wind energy adoption rises, the reallocation of traditional energy resources like fossil fuels impacts global trade patterns and economic stability. This evolution highlights the importance of foresight and adaptability in resource management.

Within the intricate fabric of economic systems, network effects intensify the impact of exponential growth on resource distribution. As networks—be they social, technological, or economic—expand, they generate value that increases disproportionately with their size. This can create self-reinforcing cycles, where the benefits of being part of a larger network attract more participants, further enhancing its value. In global trade, network effects can lead to concentrated resource allocation, granting dominant players significant advantages. Understanding these dynamics is essential for policymakers and businesses striving to create more equitable economic systems.

Chaos theory adds another layer of complexity to the relationship between exponential growth and resource allocation. While exponential growth can be somewhat predictable, the inherent unpredictability of chaotic systems poses challenges for long-term planning. Minor shifts in market conditions or consumer behavior can trigger substantial fluctuations in resource distribution, similar to the butterfly effect. This unpredictability necessitates robust modeling techniques to anticipate disruptions and develop resilient strategies. Applying chaos theory in economic contexts provides valuable insights into managing uncertainty and optimizing resource distribution during rapid growth.

Addressing the challenges of exponential growth in resource allocation requires actionable strategies. Cross-sector collaboration and fostering innovation can help build adaptable systems that respond to rapid changes. Investing in education and research to grasp the nuances of exponential growth is crucial. Encouraging critical thinking about balancing growth with sustainability is vital for future progress. By adopting these approaches, stakeholders can harness exponential growth's power to drive positive change, ensuring resources are distributed in ways that bolster economic resilience and social equity.

Network Effects in Global Trade Systems

The influence of network effects plays a crucial role in shaping global trade dynamics, revolutionizing the way nations and corporations engage in cross-border commerce. These effects occur when the value of a product or service grows as more individuals utilize it, setting off a self-reinforcing cycle that can spur rapid expansion and broad acceptance. In international trade, network effects are evident in the spread of digital marketplaces and platforms that enable effortless interactions between global buyers and sellers. E-commerce titans like Alibaba and Amazon exemplify this by harnessing extensive networks to link millions of consumers and vendors, thereby fostering economic integration and boosting efficiency. These platforms showcase how network effects can speed up the distribution of goods and services, lower transaction costs, and provide competitive advantages on a worldwide scale.

The complex mesh of global trade networks extends beyond modern innovations, tracing back to historical trade routes such as the Silk Road, which facilitated the exchange of goods, culture, and ideas across continents. Today, digital infrastructure fulfills a similar role by connecting financial markets, supply chains, and communication systems, highlighting the significance of network effects in economic advancement. Research indicates that nations with strong digital networks witness accelerated GDP growth, as they can more effectively access global markets. This connectivity also allows smaller economies to exceed expectations by engaging in international supply chains, demonstrating the democratizing potential of network-driven trade systems.

Central to the network effects in trade is the concept of positive feedback, where initial advantages can escalate into significant market dominance. This phenomenon is visible in the logistics and shipping sectors, where companies like Maersk and FedEx have established extensive networks ensuring efficient and dependable service. Such dominance poses a barrier to entry for new players, as the established network of incumbents becomes a formidable asset. However, this raises concerns about market competition and the potential for monopolistic behavior. Policymakers and economists must address these challenges to ensure network effects benefit a wide range of participants rather than concentrating power among a few.

In the rapidly evolving economic landscape, blockchain technology offers a novel means to harness network effects. By facilitating decentralized, transparent, and secure transactions, blockchain could transform the operation of global trade networks. This technology enables the creation of trustless environments where participants can engage in trade without traditional intermediaries. As blockchain adoption increases, its network effects could lead to more inclusive and resilient trade systems, reducing dependence on centralized authorities and enhancing the security of international transactions. This potential shift towards decentralized networks presents exciting opportunities for innovation and disruption in global trade.

As we examine the transformative potential of network effects in global trade systems, it is crucial to consider practical implications for businesses and policymakers. Companies can leverage these effects by strategically expanding their networks and forming partnerships that enhance connectivity and value creation. Policymakers, meanwhile, must develop regulations that balance the benefits of network-driven growth with safeguards against anti-competitive practices. By understanding and harnessing the complex dynamics of network effects, stakeholders can navigate the challenges and opportunities of an increasingly interconnected global economy, paving the way for sustainable and inclusive growth.

Chaos Theory and Predictability in Market Dynamics

The complex dance of market dynamics becomes clearer when examined through chaos theory, a discipline that deciphers the seeming randomness of complex systems to uncover hidden patterns. Often deemed unpredictable, market behavior can be better understood by acknowledging its sensitivity to initial conditions—a cornerstone of chaos theory. This sensitivity suggests that minor, seemingly trivial changes in market variables can trigger substantial shifts in economic trends over time. Such insights question the traditional belief in market efficiency, emphasizing instead the nuanced interplay between initial conditions and market forces. For instance, the butterfly effect in financial markets implies that small geopolitical events or policy shifts can lead to major economic changes, encouraging investors to consider broader impact assessments beyond conventional metrics.

Amidst the apparent turmoil, chaos theory offers a framework for identifying and predicting potential patterns within market fluctuations. By adopting this perspective, analysts and investors can develop more resilient strategies that account for the non-linear nature of market movements. Advanced computational models, based on chaos theory, allow for the simulation of market scenarios to pinpoint emerging trends and potential disruptions. These models, rooted in chaos principles, provide a valuable tool for navigating the volatile and often unpredictable global markets. Consequently, financial institutions are increasingly integrating chaos-based algorithms into their trading systems to gain a competitive advantage by predicting market behaviors that might otherwise appear erratic.

A practical grasp of chaos theory in market dynamics also extends to risk management. By recognizing the inherent unpredictability and interconnectedness of financial systems, risk managers can craft strategies that are more adaptive and resilient. Such strategies often involve stress-testing portfolios against a broad range of scenarios, including those that might initially seem unlikely. This approach not only enhances preparedness for unexpected market events but also fosters a proactive stance in managing financial risks. Investors armed with this knowledge can make more informed decisions, balancing profit pursuit with a comprehensive understanding of potential vulnerabilities within their portfolios.

In recent years, a growing body of research has explored the application of chaos theory to high-frequency trading, where rapid trade execution often magnifies the chaotic characteristics of markets. Studies reveal that high-frequency trading can create feedback loops that intensify market volatility, complicating the predictability of financial systems. This has prompted regulatory bodies to consider new frameworks to address the risks associated with such trading practices, aiming to mitigate systemic disruptions. These developments underscore the need for continued exploration of chaos theory's implications in today's fast-evolving market landscape, encouraging a blend of innovation and caution.

As readers contemplate the transformative impact of chaos theory on market dynamics, they are invited to consider its broader implications. What might the future hold for financial systems if more stakeholders embraced insights from chaos theory? How can individuals and organizations leverage this knowledge to promote more sustainable and equitable economic systems? By engaging with these questions, readers can cultivate a deeper appreciation for the intricate tapestry of market behavior and consider how their decisions might contribute to a more resilient economic future.

Cosmic Energy Transfer

At the heart of our universe, a vast and intricate exchange of power unfolds, shaping galaxies, stars, and ultimately, life. This intricate network is far from a collection of isolated phenomena; it is a harmonious blend of interwoven processes that span the cosmos, reverberating through the very fabric of space-time. As we begin this journey, we gaze into the cosmos, observing the smallest particles as they participate in a graceful dance of energy transfer, steered by the enigmatic principles of quantum mechanics. This microscopic choreography sets the stage for larger cosmic movements, illuminating how the universe manages energy distribution across vast distances.

The exploration deepens as we examine the role of electromagnetic waves, those unseen carriers that transmit energy through the void, revealing the universe's grand architecture. Their constant journey links distant celestial bodies, crafting a vibrant tapestry of energy exchange. As we delve further, the subtle influence of gravitational dynamics becomes evident, directing the flow of energy within cosmic structures, its invisible force shaping the evolution of galaxies and star systems. Within this cosmic arena, the enigmatic presence of dark matter emerges, its mysterious impact altering energy dynamics in ways that challenge our comprehension. Through this exploration, we perceive the universe not as a static realm but as a dynamic, interconnected energy system, a vibrant testament to the universal patterns at play.

Quantum Mechanics and Energy Exchange in the Cosmos

The complex ballet of energy interchange across the universe is illuminated through quantum mechanics, a theory that reshapes our grasp of reality. At its core, quantum mechanics introduces the principle of uncertainty, allowing particles to exist in multiple states at once through superposition. In this mysterious domain, energy is not simply transferred; it is a vibrant interaction of chances and potentialities. Quantum entanglement exemplifies this complexity, as particles remain linked over vast distances, instantly affecting each other's conditions. Such phenomena reveal that energy exchange at the quantum scale is a non-linear network of connections that challenges traditional logic.

Recent breakthroughs in quantum studies unveil the processes of energy transfer across the universe. Advances in quantum field theory demonstrate how virtual particles briefly emerge and vanish, facilitating energy exchanges that ripple through the cosmic framework. These fleeting particles are crucial in events like Hawking radiation, where black holes emit energy due to quantum effects at their boundaries. This once-theoretical process has gained empirical validation through advanced astronomical research, showing how quantum mechanics governs energy transfers even in the universe's most extreme conditions.

As we delve into the vast implications of quantum mechanics on cosmic energy interchange, it is vital to consider its practical uses. The emerging field of quantum computing is leveraging these principles to revolutionize energy efficiency, promising significant technological and resource-saving advancements. Quantum algorithms are crafted to optimize energy use and minimize computational demands, already proving their ability to solve complex problems with remarkable speed and precision. This not only underscores the practical applications of quantum mechanics but also hints at a future where our understanding of cosmic energy fosters substantial progress in sustainability and efficiency on Earth.

Quantum mechanics' unconventional nature also invites contemplation of the philosophical aspects of energy exchange in the cosmos. In a universe where particles can affect each other instantaneously across great distances, what implications does this have for our understanding of causality and interconnectedness? This perspective encourages a rethinking of reality, urging us to view the universe as a holographic entity where each component mirrors the entire system. By adopting this viewpoint, we open our minds to new paradigms, where energy exchange becomes not just a physical interaction but a reflection of a deeper, connected existence.

Insights into quantum energy transfer inspire us to apply these ideas at both micro and macro scales. Whether in advancing quantum technologies or unraveling the universe's fundamental laws, recognizing the quantum nature of energy transfer encourages innovative solutions to global challenges. By adopting a mindset that sees energy exchange as part of a vast, interconnected web, we can develop strategies that leverage these principles for global transformation. This perspective not only deepens our understanding of the universe but also empowers us to use these insights for the betterment of our world.

The Role of Electromagnetic Waves in Cosmic Energy Distribution

Electromagnetic waves are the universe's silent envoys, carrying energy across the vast reaches of space in a complex yet graceful manner. These waves, comprising oscillating electric and magnetic fields, are fundamental to the distribution of energy throughout the cosmos, bridging the tiniest quantum interactions and the largest cosmic events. Central to this process is the interaction between photons and matter, a fundamental exchange that determines how energy is absorbed, emitted, and transformed as it travels through the universe. Recent breakthroughs in astrophysics have uncovered the intricate behaviors of these waves, showcasing their role in phenomena like the cosmic microwave background radiation and the spectral emissions from distant galaxies. These findings not only enhance our understanding of universal energy dynamics but also offer insights into the universe's origins and development.

The exploration of electromagnetic waves in the universe has been transformed by advanced telescopes and space probes. These tools allow scientists to look beyond visible light and examine the entire electromagnetic spectrum. Each segment, from radio waves to gamma rays, reveals different aspects of cosmic energy dynamics. For example, infrared observations have illuminated the hidden processes of star formation, penetrating dust clouds that block visible light. Meanwhile, X-ray and gamma-ray studies have provided crucial insights into high-energy events like black hole accretion disks and supernova explosions, broadening our understanding of energy redistribution in the universe's most extreme environments. These discoveries highlight the essential role of electromagnetic waves in shaping the universe, providing a wealth of data that continues to drive scientific exploration.

Quantum mechanics adds another layer of complexity to our understanding of electromagnetic energy transfer. The wave-particle duality of photons allows them to act both like waves and particles, interacting with matter in ways governed by probability. This duality is central to phenomena such as quantum tunneling, where particles cross barriers they seemingly shouldn't, altering energy distributions in quantum systems. These principles not only explain the behavior of electromagnetic waves at the atomic level but also offer a framework for understanding larger cosmic processes. By embracing the probabilistic nature of quantum mechanics, researchers are beginning to unravel how electromagnetic energy influences cosmic structures, offering new perspectives on everything from galaxy formation to the behavior of dark energy.

Electromagnetic waves also interact with gravitational fields, adding depth to their role in the cosmic energy network. Gravity, often seen as merely an attractive force, interacts with electromagnetic waves to create a more intricate narrative. According to Einstein's general relativity, the curvature of spacetime affects the path of light and other electromagnetic radiation, leading to phenomena like gravitational lensing. This bending of light not only changes our view of distant cosmic objects but also acts as a natural magnifying glass, allowing us to see regions of the universe that would otherwise remain hidden. The interplay between gravitational dynamics and electromagnetic propagation reveals the elegant forces governing the cosmos, offering profound insights into the nature of reality.

As we examine the role of electromagnetic waves in the universe, intriguing questions emerge. How might these waves affect the potential for life beyond Earth, given their role in energy transfer? Could our understanding inspire new technologies for harnessing energy on Earth? Exploring these questions encourages us to consider the broader implications of cosmic energy dynamics, underscoring the interconnectedness of all systems, from the subatomic to the celestial. By contemplating the many ways electromagnetic waves shape the universe, we gain a deeper appreciation of the cosmos and our place within it, fostering a mindset that values scientific discovery and the transformative power of knowledge.

Gravitational Dynamics and Their Impact on Energy Transfer

In the grand theater of the universe, gravity serves as a pivotal force, guiding the dance of planets, stars, and galaxies, while managing the distribution of power across the cosmos. This fundamental force dictates interactions among celestial entities, establishing intricate systems that regulate the movement of energy. Central to this is the concept that mass distorts space-time, forming gravitational wells that steer the path and vitality of nearby objects. This elegant mechanism determines planetary orbits around stars and the spiral formations of galaxies, highlighting the significant influence of gravitational dynamics on universal energy distribution.

Astrophysics has recently unveiled nuanced aspects of gravitational interactions and their role in energy transfer across the universe. Gravitational waves, as forecasted by Einstein and observed by facilities like LIGO, provide a novel perspective on these interactions. As waves traverse the universe, they carry energy, offering insights into dramatic events such as black hole mergers and neutron star collisions. Through studying these phenomena, scientists not only deepen their understanding of universal energy processes but also refine knowledge of gravitational dynamics, paving the way for groundbreaking research and discoveries.

Gravitational lensing, another fascinating aspect of gravitational dynamics, offers a unique view of energy transfer in the universe. Massive objects like galaxy clusters bend the light from distant sources, acting as natural magnifiers that expose otherwise hidden cosmic regions. These lenses provide valuable opportunities to examine the distribution of both visible and dark matter, shedding light on their effects on energy movement across vast scales. By leveraging insights from gravitational lensing, astronomers can precisely map the universe's energy landscape, bridging the gap between theoretical predictions and observations.

Consider the mysterious interplay between dark matter and gravitational dynamics, a frontier challenging our understanding of the cosmos. Although unseen, dark matter exerts a gravitational pull that shapes galaxies and clusters, influencing the flow of energy through the universe. Its interaction with ordinary matter and gravitational forces creates a cosmic framework, guiding the formation and evolution of large-scale structures. Exploring this complex relationship raises questions about the true nature of dark matter and its role in the grand scheme of energy transfer, motivating researchers to devise innovative models and technologies to uncover its mysteries.

ENERGY FLOW SYSTEMS   201

For those eager to apply gravitational dynamics beyond theoretical realms, practical lessons abound. By examining how gravity affects energy movement on a cosmic scale, parallels can be drawn to systems closer to home, like energy distribution within ecosystems or economic frameworks on Earth. Understanding these dynamics fosters a holistic approach to energy management, highlighting system interconnectivity and the potential for small changes to resonate across extensive networks. Whether among the stars or within human society, gravitational dynamics offer profound insights for optimizing energy flows in a complex, interconnected world.

Dark Matter and Its Influence on Cosmic Energy Flow

The enigmatic nature of dark matter is a captivating puzzle within the universe, influencing the flow of power in ways that often elude straightforward understanding. Invisible to our eyes, dark matter neither emits nor absorbs light; however, its gravitational presence is evident throughout the cosmos. This elusive component accounts for a substantial portion of the universe's mass, guiding the development and behavior of galaxies. By exerting gravitational influence, dark matter weaves the cosmic web, affecting how energy moves through the vastness of space. Its existence is suggested by gravitational lensing, where the light from distant galaxies bends, exposing the hidden gravitational force of this unseen substance.

Recent studies into dark matter's role in universal energy dynamics have revealed intriguing theories. One hypothesis suggests that dark matter functions as an invisible framework, directing the distribution and movement of baryonic matter—the ordinary matter that forms stars and planets. This framework effect implies that dark matter's gravitational pull channels power into specific regions, encouraging galaxy formation and their dynamic interactions. Moreover, simulations have shown how dark matter's gravitational connections with visible matter can facilitate energy transfer across immense cosmic distances, offering a comprehensive model for understanding large-scale structures.

Beyond gravitational effects, dark matter's interaction with cosmic energy dynamics might extend to subtler domains. Cutting-edge research explores the potential for dark matter particles to interact weakly with ordinary matter, affecting electromagnetic fields and possibly engaging in energy exchange at a quantum level. Although these interactions remain speculative, they open new avenues for exploring how dark matter could subtly influence cosmic energy dynamics, potentially uncovering the universe's hidden mechanisms. By examining these unconventional avenues, researchers might gain deeper insights into the cosmic energy landscape.

The quest to understand dark matter's impact on cosmic energy dynamics also invites exploration of alternative models and theories. Some scientists propose changes to the standard model of particle physics, suggesting new particles or forces that could explain dark matter's mysterious nature. These theories challenge conventional thinking and encourage a reevaluation of energy dynamics in the universe. By embracing diverse perspectives, the scientific community can refine its understanding of how dark matter orchestrates energy movement on a grand scale, fostering a more nuanced appreciation of cosmic phenomena.

Dark matter's impact on cosmic energy dynamics inspires curiosity and exploration, beckoning us to consider its potential applications and implications. While tangible applications remain speculative, insights from studying dark matter could inspire novel technologies or methods for energy management on Earth. By comparing cosmic energy dynamics with terrestrial systems, we might discover principles that can be used to optimize energy distribution, leading to more sustainable practices. As we continue to unravel the mysteries of dark matter, the knowledge we gain could illuminate new pathways for addressing energy challenges on our planet, fostering a more harmonious relationship with the universe's vast energy networks.

Energy dynamics illustrate the intricate interplay of forces that fuel and transform entities across various scales. At the cellular level, energy is precisely harnessed and used to sustain life, while in the vast networks of economic systems, resources circulate to spur growth and innovation. Even on a universal scale, energy transfer shapes the formation and evolution of galaxies, stars, and the cosmos, highlighting the profound connections among all things. These systems emphasize the central theme of this book: the universal patterns that bridge the small and large scales of existence. By identifying these patterns, we gain the power to influence outcomes within our own spheres, whether through optimizing personal energy use or supporting sustainable resource distribution. As we stand ready for further exploration, the potential for positive impact by understanding and harnessing these energy dynamics becomes increasingly clear. What small steps can we take today to ignite meaningful change tomorrow? The next chapter invites readers to delve deeper into additional patterns that shape our world, continuing this journey of discovery and transformation.

Chapter Twelve
Pattern Recognition And Control

Imagine a single ripple spreading across a pond, subtly altering the path of drifting leaves and the mirrored sky. This simple movement illustrates a profound truth: recognizing and interpreting patterns can reshape our understanding of the world. This chapter delves into the realm of pattern recognition and control, inviting you to discover how identifying these subtle shifts can lead to transformative insights and significant changes.

In the dynamic tapestry of existence, identifying patterns transcends mere observation—it becomes a powerful tool for intervention and prediction. Picture having the insight to spot a crucial moment in a complex system where a minor tweak could lead to monumental change. From the intricate dance of molecules within a cell to the grand forces shaping global economies, the ability to identify these pivotal points empowers us to guide systems toward desired outcomes. This chapter unravels the art and science of detecting these key moments, enabling us to channel the currents of change.

As we venture deeper, we enter the captivating world of predictive modeling, where mathematical precision meets real-world application. With the right models, we can foresee future trends, prepare for impending challenges, and seize emerging opportunities. These models are not just theoretical; they unlock future possibilities, guiding us on paths toward innovation and sustainability. Throughout this chapter, the interplay of pattern recognition, strategic intervention, and predictive modeling will unveil new horizons for creating meaningful impacts, echoing the book's central theme of small actions driving global transformation.

Mathematical Pattern Detection

In a world inundated with endless streams of information, how do we distinguish structure from chaos? Amidst the relentless flow of data and the dominance of complexity, the skill to recognize and interpret patterns becomes not just advantageous but revolutionary. This ability forms the essence of understanding systems, whether they are as straightforward as the movement of celestial bodies or as unpredictable as the fluctuations in financial markets. Recognizing structures allows us to uncover the universe's hidden frameworks, offering insights that drive innovation and informed decision-making. It empowers us to detect meaningful signals within the noise, enabling us to predict, adjust, and even mold future possibilities.

Harnessing this potent capability demands a blend of mathematical skill and technological advancement. As we delve into the nuances of pattern recognition, we discover a rich tapestry of methods, from statistical approaches that unearth regularities amid randomness to dynamic machine learning algorithms that adapt to evolving data environments. Advanced signal processing and the emerging potential of quantum computing further expand the horizons of complex system analysis. Together, these tools create a unified framework for recognizing and influencing the systems we navigate, setting the stage for a deeper exploration into our capacity to shape the world around us.

Identifying Patterns in Noise Using Statistical Techniques

In the delicate task of discerning order within chaos, statistical methods play an indispensable role. Understanding noise not as disorder but as a potential treasure trove of hidden patterns allows us to extract significant signals. Techniques such as Fourier transforms, autocorrelation, and principal component analysis shift our view of randomness into a structured exploration for insight. These approaches dissect complex datasets, unveiling structures and correlations that are otherwise concealed. Researchers in fields from astrophysics to genomics leverage these tools to uncover phenomena obscured by randomness.

As we enter an era dominated by data, machine learning algorithms have drastically changed our approach to pattern recognition. While traditional statistics provide a solid foundation, machine learning offers a dynamic framework for learning from continually changing data streams. Neural networks and clustering algorithms enable machines to autonomously identify intricate patterns, adapting swiftly to new information. This transformation is evident in industries like finance and healthcare, where predictive analytics now forecast market trends and diagnose medical conditions with remarkable precision. These algorithms, refined through feedback, highlight the potential for machines to enhance human capability in understanding complex systems.

Advanced signal processing marks a frontier in analyzing multifaceted systems, where complex data resembles a symphony of interwoven signals. Techniques such as wavelet transforms and Hilbert-Huang transforms break down these signals, revealing their temporal and frequency characteristics. This allows for detailed examination of phenomena, unraveling the intricacies of seismic activities, brain wave patterns, and even linguistic subtleties. Such methods are invaluable in fields demanding precision, providing insights that guide interventions and innovations. By combining statistical techniques with signal processing, researchers achieve a comprehensive view of patterns hidden within noise.

The emergence of quantum computing represents a paradigm shift in pattern detection, offering unmatched processing power to solve previously insurmountable problems. Unlike classical computing's sequential operations, quantum computing utilizes superposition and entanglement to explore numerous possibilities simultaneously. This capability dramatically speeds up complex computations, identifying patterns in vast datasets at unprecedented speeds. Though still in its early stages, quantum computing promises to revolutionize fields from cryptography to drug discovery, where rapid pattern detection could lead to groundbreaking advancements.

Considering the future of pattern recognition necessitates addressing the ethical implications of these powerful technologies. As we develop more sophisticated methods for extracting information from noise, concerns about privacy, data ownership, and potential misuse arise. Balancing innovation with responsibility is crucial, ensuring our tools are used for the collective good. Encouraging interdisciplinary dialogue and fostering transparency will be vital as we navigate this complex landscape. By doing so, we not only harness the potential of pattern recognition but also ensure it aligns with broader societal values.

Machine Learning Algorithms for Dynamic Pattern Recognition

In the dynamic field of pattern recognition, machine learning algorithms have unleashed significant advancements. These algorithms autonomously learn and identify intricate structures within data, transforming how we interpret and interact with systems. Their core strength lies in their adaptive response to new information, uncovering arrangements that static analysis might miss. Consider neural networks: these systems imitate the brain's architecture, adjusting their internal parameters through techniques like backpropagation to enhance their pattern recognition capabilities with each iteration. This adaptability broadens their use across various domains, from real-time language translation to predictive maintenance in industrial applications.

Among these adaptable systems, convolutional neural networks (CNNs) excel at recognizing spatial hierarchies in data. Their layered design allows them to identify complex features within images, converting raw pixels into understandable classifications. Their utility extends beyond visual data, embracing temporal sequences like audio signals and financial trends. Here, recurrent neural networks (RNNs) and their advanced versions, long short-term memory networks (LSTMs), effectively capture time-dependent relationships. These advancements pave the way for innovations in autonomous driving and stock market forecasting, where grasping temporal dynamics is crucial.

Reinforcement learning propels the evolution of machine learning algorithms further. In this paradigm, agents refine their decision-making by interacting with their environment and receiving feedback. Through trial and error, they optimize strategies, akin to human learning. This approach has been pivotal in creating systems capable of mastering complex games like Go and chess, with applications extending well beyond recreational contexts. In healthcare, reinforcement learning aids in tailoring treatment plans, adapting strategies based on patient responses to enhance therapeutic outcomes.

Bridging theory and practical application, researchers continue to explore innovative architectures and training methods. Unsupervised learning, which discerns patterns without labeled data, opens new discovery avenues in unexplored datasets. Meanwhile, transfer learning allows models trained on one task to be repurposed for another, conserving computational resources and enhancing efficiency. This is particularly beneficial in fields with limited data, where leveraging pre-existing knowledge accelerates progress.

As machine learning algorithms evolve, they prompt reflection on their broader implications. Ethical considerations, such as training data bias and decision-making transparency, demand attention to ensure responsible use. Provocative questions emerge: How can systems be designed to recognize and understand the significance of patterns? What impact will these algorithms have on industries and society? By engaging with these inquiries, readers can contribute to the ongoing dialogue, playing an active role in shaping a future where machine learning drives meaningful change.

Advanced Signal Processing for Complex System Analysis

In the domain of advanced signal processing, mastering the art of analyzing and deciphering intricate systems is essential for revealing the complex patterns that shape our world. This field focuses on the intricate manipulation and assessment of signals that often seem chaotic, using sophisticated mathematical tools to uncover hidden structures. Central to this process is the Fourier Transform, which breaks signals into their basic frequencies, allowing the detection of patterns that might otherwise remain obscured. The applications of these techniques extend beyond conventional fields, reaching into neuroscience, where analyzing brainwave patterns aids in understanding cognitive processes, and astrophysics, where they help map cosmic radiation signatures to unlock the universe's secrets.

The advent of innovative methodologies, such as wavelet transforms, has advanced signal processing into new realms. Unlike Fourier analysis, wavelets provide multi-resolution analysis, offering a detailed view of signals by capturing both time and frequency information simultaneously. This method is particularly effective in areas like seismology, where identifying subtle shifts in seismic waves can help predict earthquakes. By breaking down signals into time-frequency representations, wavelets enable scientists to identify patterns with exceptional precision, providing insights crucial for both theoretical advancements and practical applications. This detailed analysis highlights the versatility of advanced signal processing techniques, ensuring their prominence in scientific exploration.

Quantum computing introduces a revolutionary era for signal processing, promising exponential improvements that could transform pattern detection in complex systems. Quantum algorithms, utilizing qubits and superposition, possess the potential to solve problems currently unsolvable by classical computers. This capability is especially relevant in processing vast datasets, such as those from global climate models or genomic sequences. By exploiting quantum algorithms, researchers can uncover patterns and correlations on an unprecedented scale, offering a clearer understanding of dynamic systems and their interactions. The convergence of quantum computing and signal processing represents a paradigm shift that could redefine our approach to pattern recognition and control.

The real-world implications of advanced signal processing are extensive, influencing critical areas like cybersecurity and financial markets. In cybersecurity, anomaly detection algorithms analyze network traffic to spot patterns indicative of potential threats, ensuring proactive defense strategies. Similarly, in financial markets, signal processing assists in forecasting stock trends by examining historical data patterns, providing investors with strategic insights. These applications highlight the significance of signal processing in managing complex systems, where detecting subtle patterns can mean the difference between success and failure. As these techniques continue to evolve, their impact on strategic decision-making across industries will only grow.

To fully harness the potential of advanced signal processing, ongoing research and development are essential. Integrating interdisciplinary approaches that combine artificial intelligence, machine learning, and quantum mechanics can further enhance signal analysis capabilities. By encouraging collaboration across scientific fields, we can push the boundaries of what is possible, empowering individuals and organizations to utilize patterns innovatively. This pursuit not only enriches our understanding of the world but also equips us with the tools to drive meaningful change, fostering a proactive approach to the complex challenges ahead.

Quantum Computing's Role in Enhanced Pattern Detection

Quantum computing is revolutionizing how we detect patterns, offering new capabilities that exceed the boundaries of traditional computing. Central to this innovation is the qubit, which stands apart from the classical binary bit. While bits are either 0 or 1, qubits can exist in states of 0, 1, or any combination of both, enabling quantum computers to process immense data sets simultaneously. This unique feature allows for the swift analysis of intricate systems where patterns are buried beneath layers of noise. Quantum computing's ability to uncover complex configurations, often hidden from classical systems, is paving the way for advancements in cryptography, material science, and financial modeling.

A compelling application of quantum computing in pattern recognition is its enhancement of machine learning algorithms. Quantum machine learning (QML) combines quantum algorithms with traditional frameworks, drastically speeding up tasks such as classification, clustering, and regression. Quantum-enhanced support vector machines and neural networks can quickly identify structures within data sets, using fewer resources than conventional methods. This boost in efficiency is particularly useful for real-time data analysis in dynamic settings, like automated trading systems or fraud detection, where rapid and precise pattern recognition is crucial.

Signal processing, too, benefits significantly from quantum computing, which transforms or analyzes signals to extract vital information. Quantum algorithms, including the quantum Fourier transform, markedly increase the efficiency of signal processing. These algorithms can break down complex signals into their basic frequencies with extraordinary precision, helping reveal arrangements that were previously hidden. Such capabilities are essential in fields like telecommunications, where identifying patterns amidst noise can substantially improve data transmission quality and reliability.

Theoretical exploration also sees the influence of quantum computing, challenging established paradigms and offering new insights into long-standing problems. An exciting research area is the use of quantum algorithms for pattern recognition in biological systems. By simulating complex biological processes like protein folding or genetic mutations, quantum computers can discover patterns that might lead to breakthroughs in personalized medicine and bioinformatics. These findings not only enhance scientific knowledge but also open pathways to novel treatments and more efficient drug discovery.

As quantum computing progresses, its influence on pattern detection will only grow. Researchers are exploring hybrid approaches that blend classical and quantum computing, using the strengths of both to tackle challenges neither could solve alone. This collaboration promises to improve our ability to identify patterns in increasingly complex networks, from climate modeling to astrophysics. By encouraging interdisciplinary partnerships, quantum computing's potential can be fully harnessed, transforming our understanding and interaction with the world. These advancements invite readers to consider integrating quantum technologies into their own fields, expanding the boundaries of what is achievable in pattern detection and control.

System Intervention Points

In recent years, we have witnessed a growing awareness of the pivotal moments within intricate networks where precise, minor actions can generate significant changes. These intervention points act as concealed levers, waiting to be discovered and strategically used to drive transformation. By pinpointing and leveraging these opportunities, we can guide complex frameworks—be they environmental, societal, or technological—toward desired outcomes. This exploration transcends academic inquiry; it is an invitation to harness the subtle yet potent forces shaping our world. As we refine our ability to recognize these critical junctures, we gain the capacity to influence outcomes, steering systems from disorder to balance and resilience.

The journey into uncovering intervention points starts with learning to identify these hidden prospects amid the complexities of system dynamics. Spotting these leverage points requires a sharp eye for interactions that can be magnified to initiate change. Once identified, assessing the sensitivity and durability of these points becomes vital, guiding us in crafting interventions that are impactful and sustainable. Designing strategic measures demands not only insight but also creativity, blending the art and science of achieving optimal impact. Advanced methodologies enable us to disrupt and reshape systems, challenging the limits of possibility. This nuanced approach embodies the essence of recognizing and controlling patterns, equipping readers with the tools to engage with the world in transformative and positive ways.

Identifying Leverage Points in Complex Systems

Understanding leverage points within intricate systems demands insight into the subtle interconnections and concealed forces shaping their behavior. These points, highlighted by systems theorists, are pivotal locations where minimal adjustments can instigate significant transformations. They often lie hidden in system architecture, regulations, or information flow, rather than in overt elements like resources or outputs. Recent research underscores their importance, showing that effective interventions usually involve modifying feedback loops, altering mindsets, or redefining system goals. By pinpointing these leverage points, one can guide a system towards better outcomes with minimal effort, much like steering a ship by adjusting its rudder.

Determining which system elements have the most transformative potential is challenging. This necessitates a thorough examination of the system's framework to identify influential nodes and connections. Network theory offers valuable tools for this purpose, enabling the mapping and analysis of relationships among system components. In social systems, for example, key influencers or opinion leaders might serve as leverage points, where impacting a few can lead to broad behavioral changes. Similarly, in ecological systems, keystone species or critical habitats might act as leverage points for preserving biodiversity. This strategy not only helps focus efforts but also anticipates the ripple effects of interventions, ensuring changes unfold as intended.

Grasping a system's resilience and sensitivity is crucial for effectively using leverage points. Systems with high resilience may require more substantial interventions for noticeable impact, while highly sensitive systems might react profoundly to minor changes. This resilience-sensitivity spectrum is key in selecting and prioritizing leverage points. Advances in system dynamics modeling now allow simulations of potential interventions, letting researchers and practitioners test and refine strategies without real-world risks. These computational tools offer a safe space for experimentation, revealing how different leverage points respond under various scenarios, ultimately aiding precise and informed decision-making.

Crafting strategic interventions at leverage points requires creativity and foresight, balancing immediate benefits with long-term sustainability. It's essential to consider the cultural, ethical, and social aspects of these interventions, as they can deeply affect acceptance and success. In public health, for instance, interventions targeting social norms or promoting behavior change can act as leverage points for improving community health outcomes. A holistic perspective that integrates diverse viewpoints and disciplines can enrich intervention strategies, fostering approaches that are innovative and inclusive. This multidisciplinary approach not only strengthens solutions but also helps mitigate unintended consequences that might arise from narrowly focused actions.

Identifying and using leverage points in complex systems is both an art and a science. It demands analytical rigor, intuition, and a readiness to embrace uncertainty. Practitioners should nurture curiosity and open-mindedness, constantly questioning assumptions and exploring alternative paths. By staying attuned to the latest in system dynamics research and engaging in collaborative, cross-disciplinary efforts, individuals and organizations can harness leverage points to effect meaningful and lasting change. This approach turns the daunting complexity of systems into an opportunity for innovation and progress, empowering readers to become adept navigators in pursuing global transformation.

Evaluating Sensitivity and Resilience in System Dynamics

Exploring complex systems requires a keen understanding of sensitivity and resilience, which are crucial for those aiming to stabilize or influence dynamic environments. Sensitivity in this context refers to a system's reactivity to initial conditions or external inputs. A highly sensitive system can undergo major changes from minor influences, resembling the butterfly effect in chaos theory, where a small disturbance may lead to vastly different outcomes. On the other hand, resilience measures a system's capacity to absorb disruptions while maintaining core functions. Balancing these attributes helps strategists and researchers design interventions that promote desired results while minimizing potential disruptions.

Recent advancements in data science and computational modeling have transformed our ability to evaluate and predict systems' sensitivity and resilience. Techniques like agent-based modeling and machine learning algorithms enable the simulation of numerous scenarios, providing insights into how systems might react under various conditions. For example, in ecology, researchers use these models to forecast how ecosystems respond to climate change or human interference, offering vital data for conservation efforts. By simulating different scenarios, stakeholders can pinpoint influential variables and identify interventions that strengthen the system against potential threats.

Adaptive capacity, increasingly acknowledged as a key element of resilience, refers to a system's ability to adjust its functioning in response to changes, thereby upholding its core objectives. In economic systems, for instance, the ability to adapt to market fluctuations can determine a business's longevity and success. Organizations that nurture innovation, flexibility, and collaboration typically exhibit higher adaptive capacity, enabling them to thrive amid uncertainty. Promoting these traits can enhance a system's ability to withstand and recover from unexpected challenges, ensuring long-term viability.

Assessing system dynamics through the lens of sensitivity and resilience also involves considering tipping points—thresholds where a system's behavior shifts dramatically. Understanding these points is crucial for realizing how small, targeted actions can lead to significant changes, both positive and negative. Identifying and leveraging these tipping points requires a nuanced appreciation of a system's internal and external interactions. In social systems, for example, recognizing the critical mass needed for societal change can inform strategic planning and resource allocation, increasing the likelihood of transformational success.

Engaging with the dynamic interplay of sensitivity and resilience provides actionable insights for those seeking to responsibly influence systems. Practitioners should begin with comprehensive assessments to map a system's current state, identifying areas of vulnerability and strength. From there, designing interventions that bolster adaptive capacity and prepare for potential tipping points becomes feasible. Fostering collaboration across disciplines and sectors enriches the understanding and management of complex systems, creating environments where positive change can thrive. By adopting these principles, readers are equipped to navigate and shape their respective domains, driving impactful and sustainable progress.

Designing Strategic Interventions for Optimal Impact

In the complex web of dynamic frameworks, crafting strategic measures is a deliberate balance of exactness and ingenuity. Central to impactful actions is recognizing leverage points—those critical spots within a framework where small changes can trigger significant shifts. Recent research in systems theory has underscored these points as vital for driving change, demonstrating their ability to steer frameworks in preferred directions. For instance, in environmental management, focusing on essential leverage points like carbon reservoirs or biodiversity hubs can significantly bolster ecological stability and sustainability. This refined insight elevates framework intervention from simple management to an intricate art, where subtle influence leads to profound results.

A key element in devising effective measures is assessing a framework's sensitivity and resilience. Sensitivity analysis is an advanced quantitative tool that examines how input variations impact outputs. It offers a guide to understanding which parts of a framework are most vulnerable and which are robust enough to endure disturbances. A resilient framework, capable of absorbing disruptions while maintaining functionality, presents ideal conditions for strategic measures. This dual focus on sensitivity and resilience ensures actions are not only impactful but also sustainable in the long run, promoting a balance between stability and adaptability.

Designing strategic measures also involves using predictive modeling to anticipate outcomes. By simulating various scenarios, these models provide valuable insights into potential impacts, enabling practitioners to refine their strategies with foresight. In healthcare, predictive analytics can identify the most effective measures for patient care, optimizing resource allocation and improving health outcomes. This predictive approach transforms action design from reactive to proactive, allowing practitioners to foresee challenges and opportunities with remarkable precision.

In the sphere of framework transformation, advanced techniques like disruption and reconfiguration are becoming more prominent. These methods challenge conventional ideas by deliberately destabilizing existing structures to make way for innovation. This approach leverages cutting-edge research in fields like organizational change and technology adoption, where strategic disruptions have spurred breakthroughs. By embracing disruption as a transformative force, practitioners can unlock the potential for radical change, overturning conventional thinking and fostering environments conducive to innovation and progress.

As we explore strategic measures, it's vital to adopt a mindset of experimentation and iteration. The complexity of frameworks demands a flexible approach, where actions are continually assessed and refined based on real-time feedback. This iterative process fosters critical thinking and adaptive learning, ensuring strategies remain relevant in an ever-changing landscape. By cultivating a culture of curiosity and openness to diverse perspectives, practitioners can discover new avenues for impact, transcending traditional limits and paving the way for a future where small, deliberate actions create waves of global change.

Advanced Techniques for System Disruption and Transformation

In the field of complex networks, disrupting and transforming systems requires a deep understanding of advanced strategies that transcend traditional intervention methods. System disruption involves more than just dismantling existing frameworks; it strategically alters the balance to enable transformation. This demands a thorough grasp of the system's structure and the identification of key pressure points that, when targeted, can provoke substantial shifts. A prime example is ecological management, where introducing a keystone species can recalibrate an ecosystem, exemplifying significant changes through minimal initial interventions. These principles extend beyond ecology, applying to organizational change management, where a crucial policy or innovation can trigger widespread transformation.

Recent progress in computational modeling and artificial intelligence provides novel avenues for system disruption. Machine learning algorithms now predict system behavior with impressive accuracy, uncovering subtle intervention points. They simulate various scenarios in a virtual environment before real-world application, significantly reducing the risks of disruptive actions. In financial markets, for example, algorithmic trading employs predictive models to influence market trends, capitalizing on micro-fluctuations while maintaining strategic stability and growth. Applying these predictive tools in social systems could revolutionize approaches to societal challenges, offering scalable solutions to intricate problems.

Disruption often faces resistance, as systems naturally strive to maintain equilibrium. Thus, understanding a system's resilience and sensitivity becomes vital. Highly resilient systems require more strategic and innovative interventions for transformation. Here, the concept of adaptive cycles proves relevant. By aligning interventions with natural phases of growth, conservation, release, and reorganization, one can orchestrate disruptions that harmonize with a system's inherent dynamics. In urban planning, for instance, integrating green infrastructure during a city's expansion phase can promote long-term sustainability, leveraging the natural growth cycle for transformation.

Incorporating diverse perspectives and interdisciplinary approaches further enhances the toolkit for system disruption. Insights from fields like behavioral economics, neuroscience, and cultural studies can reveal unconventional strategies that traditional methodologies might overlook. These perspectives challenge entrenched viewpoints and offer fresh avenues for exploration. By considering cultural context, social norms, and behavioral patterns within a system, interventions can be tailored to resonate more deeply, increasing their effectiveness. This holistic approach is exemplified in public health campaigns that combine cultural sensitivity with scientific insight to effectively alter health behaviors on a large scale.

For practitioners aiming to implement these advanced techniques, adaptability and continuous learning are essential. The landscape of complex systems is ever-changing, requiring a mindset open to experimentation and iteration. Engaging with the latest research, participating in interdisciplinary dialogues, and fostering a culture of innovation within organizations are crucial steps. Practitioners can conduct small-scale pilot projects to refine strategies before broader application, ensuring interventions are both impactful and sustainable. By cultivating an environment where novel ideas are nurtured and diverse perspectives are welcomed, leaders can drive transformative change, utilizing disruption to pave the way for a more resilient future.

Predictive Modeling Applications

Predictive modeling bridges the worlds of mathematics and creativity, transforming abstract data into concrete predictions that influence our perception of reality. As we delve into the realm of prediction, we navigate a complex web of interconnected principles, spanning from the everyday to the extraordinary. These models guide decisions daily, from predicting traffic patterns to anticipating economic shifts. They offer us a glimpse into a future shaped by intricate networks, where even minor adjustments can cascade into significant changes. The true marvel is in identifying recurring themes across various fields, allowing us to foresee and sometimes influence outcomes. This capability not only deepens our understanding of intricate systems but also empowers us to leverage these insights for meaningful change.

In exploring the subtleties of predictive modeling, we encounter the significant role of feedback loops—those cyclical influences that drive dynamic predictions. In social contexts, network effects illuminate collective human behavior, illustrating how ideas proliferate and movements gain traction. Chaos theory, with its unpredictable nature, further broadens our predictive tools, helping us unravel the complexities of weather and climate. At the core of complexity, evolutionary algorithms stand as formidable problem-solvers, adept at navigating expansive solution landscapes with nature-inspired finesse. Each segment builds upon the previous, crafting a narrative that enhances our ability to anticipate, adapt, and ultimately shape the future in alignment with our visions of positive transformation.

Harnessing Feedback Loops for Dynamic System Predictions

Understanding the intricate nature of feedback loops in dynamic system predictions reveals the complex interplay between cause and effect in multifaceted environments. These loops, whether they amplify or stabilize processes, are vital to how systems change over time. In predictive modeling, they offer a unique perspective to foresee shifts and trends. By analyzing the interactions among various elements within a system, we can develop models that not only forecast results but also adjust to evolving conditions. This flexibility is essential in areas like economics, where market dynamics change swiftly, and precise predictions can provide strategic benefits.

Recent strides in computational capabilities and machine learning have significantly enhanced our ability to accurately model feedback loops. Advanced algorithms now simulate numerous scenarios at once, offering a more detailed view of possible future states. For example, in ecological contexts, understanding the predator-prey relationship through feedback loops leads to more accurate conservation strategies. By observing how population variations affect an ecosystem, we can design interventions that sustain balance and avert collapse. This advanced approach moves beyond traditional static models, providing a dynamic framework that reflects the fluidity of real-world systems.

Incorporating feedback loops into predictive models also fosters a deeper understanding of system interdependencies. Take the financial sector, where feedback loops often appear in stock market movements. Analysts who grasp these loops can better anticipate market trends, making more informed decisions. By pinpointing key variables and mapping their interactions, predictive models can estimate not just the direction but also the extent of change. This knowledge equips stakeholders to craft strategies that are proactive and resilient, reducing risks in a constantly changing market.

Exploring feedback loops in predictive modeling extends beyond established fields, opening new avenues in emerging disciplines. In personalized medicine, for instance, feedback loops play a crucial role in customizing treatment plans for individual patients. By tracking how a person responds to medication, models can suggest the best dosages and treatment paths, enhancing outcomes and minimizing side effects. This patient-focused approach showcases the transformative power of feedback-driven predictions, where anticipating and adapting are key.

As we continue to refine predictive modeling through feedback loops, it's vital to consider the broader implications. These models are not just tools for forecasting; they offer insights into the underlying structure of complex systems. Engaging with this understanding equips us to influence outcomes in meaningful ways. Readers might contemplate: How could an awareness of feedback loops in their own lives lead to more informed decisions and positive changes? Embracing this exploration allows one to actively shape the future through thoughtful interaction with the systems that govern our world.

Leveraging Network Effects to Anticipate Social Movements

Network effects are instrumental in comprehending and forecasting the dynamics of social movements. These effects arise when the significance or influence of an event grows as more individuals participate, creating a cascading effect that can trigger substantial societal transformations. In the realm of social movements, the digital interconnectedness of people through platforms and communication technologies is pivotal. By exploring how these connections enhance individual actions, we can more accurately predict social change trajectories. Recent analyses have highlighted how movements like #MeToo and

Black Lives Matter effectively utilize network effects, swiftly spreading their messages and rallying global support. The digital era has redefined traditional paradigms, enabling unprecedented levels of engagement and influence.

To foresee the emergence and progression of social movements, pinpointing key influencers within networks is crucial. These influencers often act as accelerators, spreading ideas and encouraging collective action. Advanced algorithms and data analytics serve as tools to map these networks, uncovering patterns and potential intervention points. Research shows that movements frequently gain momentum through the strategic involvement of prominent individuals or groups who amplify the cause. Understanding these dynamics allows strategists to anticipate movements and design interventions to either support or mitigate their impacts, depending on desired outcomes.

Network effects extend beyond technology-driven movements. Historical examples like the civil rights movement illustrate how interconnected relationships and common goals can propel societal change. The difference today is in the speed and scope of these networks. With social media and instant communication, movements gather momentum much faster. This acceleration calls for new study methods, incorporating real-time data analysis and predictive modeling to stay ahead of trends. Researchers are increasingly focusing on the convergence of sociology, technology, and mathematics to develop comprehensive models that predict social movement dynamics.

Despite the potential for predicting social movements, challenges exist in accounting for the unpredictable human element. People are not just nodes in a network; they bring diverse emotions, motivations, and perspectives that can alter trajectories unexpectedly. Chaos theory offers insights into this unpredictability, indicating that minor changes can lead to significant, unforeseen outcomes. By integrating chaos theory principles with network analysis, more nuanced models can be constructed, accommodating the complex interplay of order and disorder in social phenomena. These models can identify early indicators of change, providing valuable foresight into social evolution dynamics.

With this knowledge, individuals and organizations can take steps to positively influence social movements. This involves understanding and actively participating in networks to achieve desired outcomes. Engaging with key influencers and recognizing the power of collective voices allows stakeholders to shape narratives and drive meaningful change. The key is to adopt a proactive approach, leveraging insights from network analysis to craft strategies aligned with broader societal goals. As our understanding of network effects evolves, so will our ability to predict and shape the future of social movements, paving the way for informed and intentional action in an increasingly connected world.

Utilizing Chaos Theory in Weather and Climate Forecasting

Exploring chaos theory reveals powerful insights into the complexities of weather and climate forecasting. Weather behaviors, as chaotic systems, are sensitive to initial conditions—a phenomenon famously known as the "butterfly effect." Minor alterations in atmospheric conditions can lead to drastically different outcomes. This inherent unpredictability challenges meteorologists but also paves the way for innovative predictive models. By embracing the stochastic nature of atmospheric dynamics, scientists can create more resilient simulations that account for a wider array of potential scenarios, enhancing forecast accuracy and reliability.

The inclusion of chaos theory transcends traditional linear models, introducing advanced algorithms capable of capturing the nonlinear interactions within the atmosphere. These algorithms, anchored in chaos mathematics, excel in identifying subtle signals amidst noise, providing a clearer vision of future weather scenarios. Recent advancements in computational power and machine learning have significantly advanced this field, enabling the synthesis of vast datasets and refining predictive models. These innovations not only enhance short-term weather forecasts but also improve long-term climate projections, offering valuable insights for policymakers and stakeholders.

Incorporating chaos theory into climate modeling requires acknowledging the feedback mechanisms within Earth's systems. Complex interactions, such as ocean currents, atmospheric pressure, and temperature gradients, often lead to emergent behaviors that are not easily explained. Recognizing these patterns and their implications allows scientists to better anticipate phenomena like El Niño or La Niña, which have wide-ranging effects on global weather systems. This understanding supports more informed decision-making in areas such as agriculture, disaster preparedness, and resource management.

Practical applications of chaos theory in forecasting extend beyond theoretical models. Forecasters are increasingly adopting ensemble forecasting, a method involving multiple simulations with slightly varied initial conditions. This technique offers a probabilistic range of outcomes, providing a more comprehensive understanding of potential future states. By presenting these possibilities, rather than a singular deterministic forecast, stakeholders can better prepare for various scenarios, reducing risks associated with extreme weather events. This strategic foresight is essential in adapting to the growing volatility induced by climate change.

Viewing chaos theory through a broader lens in weather and climate forecasting shows that embracing uncertainty is not a constraint but an opportunity. Acknowledging the complex, interdependent nature of atmospheric systems enables resilience and adaptability in responding to environmental challenges. As predictive capabilities are refined, we are not just forecasting weather; we are shaping our collective future, empowering communities to thrive amidst uncertainty. Through this perspective, chaos theory evolves from an abstract mathematical concept into a crucial tool for global change and sustainability.

Applying Evolutionary Algorithms for Complex Problem-Solving

Exploring the potential of evolutionary algorithms reveals a compelling method for tackling complex challenges, inspired by the natural processes of selection, mutation, and survival. These algorithms shine in intricate environments where conventional approaches struggle, facilitating the efficient and creative navigation of extensive solution landscapes. By progressively enhancing a collection of potential solutions and choosing the most promising ones based on set criteria, evolutionary algorithms have become powerful tools across various fields, from engineering to artificial intelligence. They are particularly useful when solutions are not immediately clear, requiring a process akin to natural evolution.

Recent breakthroughs have expanded the scope and efficiency of evolutionary algorithms. Researchers now use multi-objective optimization, where multiple criteria are optimized simultaneously, offering a more comprehensive problem-solving strategy. In aerodynamic design, for instance, these algorithms can balance minimizing drag and maximizing structural integrity. Such innovations not only improve these algorithms' performance and applicability but also pave the way for new advancements across different sectors, pushing the limits of what they can achieve.

Incorporating evolutionary algorithms into predictive modeling demonstrates their adaptability in dynamic and uncertain settings. They can adjust to new data and shifting conditions, making them invaluable for applications like financial market forecasts or adaptive robotics. A notable example is their role in developing autonomous vehicles, where they optimize navigation strategies in real-time, adapting to changing traffic and unforeseen obstacles. This adaptability highlights the transformative impact of evolutionary algorithms, enabling continuous learning and improvement in systems that require quick and responsive decision-making.

The versatility of evolutionary algorithms extends to fostering innovation in creative fields. In music and art, these algorithms can generate new compositions and visual designs by simulating the creative process. By embedding aesthetic principles or user preferences, the system can produce outputs that not only meet specific criteria but also exhibit unique and surprising qualities. This fusion of technology and creativity underscores the potential of evolutionary algorithms to break traditional boundaries, offering new forms of expression and exploration.

To fully harness the capabilities of evolutionary algorithms, practitioners must address several factors, including selecting appropriate fitness functions, balancing exploration and exploitation, and managing computational resources. Addressing these considerations can lead to the creation of more robust and efficient algorithms tailored to specific problem domains. By refining these elements, practitioners can unlock the full potential of evolutionary algorithms, driving innovation and achieving breakthroughs in areas that demand sophisticated problem-solving abilities. This approach not only deepens our understanding of these algorithms but also empowers individuals and organizations to leverage them for meaningful and impactful change.

Reflecting on our exploration of pattern recognition and control, the profound influence of identifying and manipulating structures within various networks becomes evident. This examination of mathematical pattern detection reveals how these concepts can uncover foundational arrangements across diverse mechanisms, presenting avenues for strategic action. By recognizing crucial intervention points, we can utilize these insights to drive significant change, whether it pertains to ecological harmony, economic resilience, or technological progress. The application of predictive modeling further empowers us by converting knowledge into actionable foresight, enabling us to anticipate and prepare for future challenges.

This chapter has highlighted the delicate interplay between recognizing structures and exercising control, urging readers to regard small actions as catalysts for considerable transformation. As we move forward to explore the universal principles of patterns, consider how these insights might motivate your journey toward fostering positive impact. What structures can you identify, and how might you leverage them to contribute to a broader, beneficial outcome?

Conclusion

As we draw this exploration to a close, we find ourselves with a deeper understanding of the intricate designs that shape our world. From the subtle dance of molecules to the grandeur of the cosmos, these patterns serve as a universal language that transcends both scale and context. By delving into these recurring themes, we have unearthed the mathematical principles that echo through diverse systems, demonstrating how minor actions can lead to significant transformations.

Fundamental Design Concepts

At the core of our exploration lies a recognition of fundamental design concepts that permeate the fabric of reality. These concepts, including self-replication, feedback mechanisms, and network effects, form the essential building blocks of change. They appear in various manifestations, from the spiral growth of a fern to the complex choreography of galaxies. Understanding these principles provides us with insights into the forces that drive both stability and transformation, enabling us to perceive the world through a lens of interconnected dynamics. Throughout the chapters, we have explored how these structures underpin the behaviors of complex systems, offering a framework to decode chaos and discern order. This comprehension not only enriches our intellectual pursuits but also equips us with the tools to engage more mindfully with our environment, recognizing the echoes of these universal designs in the systems we inhabit.

Multi-Level Application Methods

The journey through this book also highlights strategies for applying these concepts across different scales, empowering us to instigate change from the micro to the macro level. The multi-level application methods discussed provide a roadmap for harnessing the power of these patterns in practical ways. By drawing parallels between seemingly unrelated domains, we can apply insights from cellular behavior to social movements, or from economic cycles to ecological systems. This cross-pollination of ideas fosters innovation and resilience, enabling us to devise solutions that are both adaptive and robust. The book encourages us to think beyond conventional boundaries, urging us to embrace a systems-thinking approach that considers the ripple effects of our actions. Whether in personal endeavors or collective initiatives, these strategies inspire us to harness design principles to create positive change, reminding us that even the smallest actions can have far-reaching impacts.

Future Frontier Discoveries

As we stand at the threshold of future exploration, the potential for discovering new designs beckons with promise and excitement. The frontiers of pattern discovery are as vast as they are intriguing, offering endless possibilities for expanding our understanding of the universe. Advances in technology and data analysis open new avenues for uncovering hidden structures, challenging us to delve deeper into the complexities of the systems we study. The book invites us to embrace this sense of curiosity, encouraging us to be pioneers in the exploration of emergent phenomena and to seek out the designs that will shape the future. By remaining open to new insights and perspectives, we position ourselves at the forefront of innovation, ready to contribute to the evolving tapestry of knowledge. This forward-looking perspective not only inspires us to continue our exploration but also emboldens us to apply our newfound understanding to address the pressing challenges of our time.

Reflecting on the insights gained, we recognize the transformative power of understanding these patterns in shaping our perception of the world. The knowledge acquired from this exploration instills a profound sense of empowerment, urging us to actively participate in global change. By internalizing the lessons of fundamental design concepts, multi-level application methods, and the exciting frontiers of future discovery, we are equipped to navigate the complexities of an interconnected world with confidence and purpose. The book leaves us with a renewed sense of wonder and anticipation, inviting us to continue the journey of exploration and to apply the knowledge we have gained in meaningful ways. As we move forward, let us carry with us the realization that every action, no matter how small, has the potential to mirror and magnify change on a global scale.

Resources

Books

1. "Sync: The Emerging Science of Spontaneous Order" by Steven Strogatz - This book explores the concept of synchronization across various systems, from fireflies to human biology, offering insights into the patterns discussed in Micro Mirrors. Link to book

2. "The Selfish Gene" by Richard Dawkins - A classic exploration of evolutionary biology that provides a foundation for understanding self-replication and evolutionary algorithms. Link to book

3. "Chaos: Making a New Science" by James Gleick - An introduction to chaos theory, explaining its impact on various scientific fields and its relevance to the unpredictable patterns covered in Micro Mirrors. Link to book

4. "Ubiquity: Why Catastrophes Happen" by Mark Buchanan - This book discusses power laws and phase transitions, offering a deeper understanding of why large-scale changes often start with small triggers. Link to book

5. "The Structure of Scientific Revolutions" by Thomas S. Kuhn - Examines the patterns of scientific progress and paradigm shifts, relevant to understanding pattern recognition and control. Link to book

Websites

1. Khan Academy - Offers free courses on mathematics and science, providing foundational knowledge on many of the mathematical principles discussed in Micro Mirrors. Khan Academy

2. Coursera - Features courses on complex systems, network theory, and other topics relevant to understanding global change patterns. Coursera

3. Complexity Explorer - A resource from the Santa Fe Institute offering online courses and materials on complex systems science. Complexity Explorer

4. Wolfram Alpha - A computational knowledge engine useful for exploring mathematical patterns and principles in real-time. Wolfram Alpha

5. Edge.org - Provides thought-provoking articles and discussions from leading scientists and thinkers on topics related to patterns and change. Edge.org

Articles

1. "The Mathematics of Emergence" by Philip Ball - An article exploring how simple rules can lead to complex behaviors, relevant to network effects and emergence. Link to article

2. "Feedback Loops: How Nature Gets its Rhythms" by Carl Zimmer - Discusses the importance of feedback loops in natural systems, providing additional context for Chapter 2 of Micro Mirrors. Link to article

3. "The Science of Pattern Formation" by Philip Ball - An exploration of how patterns form in nature, applicable to the study of symmetry and breaking patterns. Link to article

4. "Power Laws in Economics and Finance" by Xavier Gabaix - Examines the role of power laws in economic contexts, offering insights into wealth distributions and market behavior. Link to article

5. "The Role of Chaos in Weather Prediction" by Edward Lorenz - A classic paper explaining how chaos theory impacts weather prediction models, paralleling discussions in Chapter 9 of Micro Mirrors. Link to

article

Tools

1. Gephi - An open-source network analysis and visualization software that helps in exploring and understanding complex networks. Gephi

2. NetLogo - A multi-agent programmable modeling environment for simulating natural and social phenomena, useful for exploring self-replicating patterns and feedback loops. NetLogo

3. MATLAB - A high-performance language for technical computing, offering capabilities for data analysis and visualization of mathematical patterns. MATLAB

4. RStudio - An integrated development environment for R, ideal for statistical computing and graphics related to pattern analysis. RStudio

5. Python (with SciPy and Matplotlib) - A versatile programming language with libraries for scientific computing and visualization, useful for exploring the mathematical principles discussed in Micro Mirrors. Python

Organizations

1. Santa Fe Institute - A research institute specializing in complex systems science, offering resources and events related to the topics in Micro Mirrors. Santa Fe Institute

2. The Complexity Science Hub Vienna - Focuses on the science of complex systems, providing collaborative research opportunities and insights. Complexity Science Hub Vienna

3. Institute of Physics - Offers resources and publications on physics-related topics, supporting the exploration of wave dynamics and quantum phenomena. Institute of Physics

4. The Royal Society - A fellowship of scientists promoting excellence in the sciences, providing access to cutting-edge research and discussions. The Royal Society

5. The Center for the Study of Complex Systems at the University of Michigan - An interdisciplinary research center offering insights into complex systems research. CSCS at UMich

These resources provide a comprehensive starting point for readers to delve deeper into the concepts covered in "Micro Mirrors." By exploring these books, websites, articles, tools, and organizations, readers can further enhance their understanding and application of the universal patterns that drive global change.

References

1.

Anderson, P. W. (1972). More is different. Science, 177(4047), 393-396. Bak, P., Tang, C., & Wiesenfeld, K. (1987). Self-organized criticality: An explanation of 1/f noise. Physical Review Letters, 59(4), 381-384. Barabási, A.-L. (2003). Linked: How everything is connected to everything else and what it means for business, science, and everyday life. Plume. Bass, F. M. (1969). A new product growth for model consumer durables. Management Science, 15(5), 215-227. Bateson, G. (1972). Steps to an ecology of mind: Collected essays in anthropology, psychiatry, evolution, and epistemology. University of Chicago Press. Casti, J. L. (1994). Complexification: Explaining a paradoxical world through the science of surprise. HarperCollins. Chandrasekhar, S. (1961). Hydrodynamic and hydromagnetic stability. Oxford University Press. Dee, D., & Ghil, M. (1984). Boolean difference equations, I: Formulation and dynamic behavior. SIAM Journal on Applied Mathematics, 44(1), 111-126. Foster, J. (2005). From simplistic to complex systems in economics. Cambridge Journal of Economics, 29(6), 873-892. Gleick, J. (1987). Chaos: Making a new science. Viking. Goldstein, J. (1999). Emergence as a construct: History and issues. Emergence, 1(1), 49-72. Haken, H. (1983). Synergetics: An introduction. Springer. Holland, J. H. (1992). Adaptation in natural and artificial systems. MIT Press. Kauffman, S. A. (1993). The origins of order: Self-organization and selection in evolution. Oxford University Press. Kitcher, P. (1982). Genes. The British Journal for the Philosophy of Science, 33(4), 337-359. Lorenz, E. N. (1963). Deterministic nonperiodic flow. Journal of the Atmospheric Sciences, 20(2), 130-141. Mandelbrot, B. B. (1983). The fractal geometry of nature. W.H. Freeman. Maturana, H. R., & Varela, F. J. (1980). Autopoiesis and cognition: The realization of the living. Reidel. Mitchell, M. (2009). Complexity: A guided tour. Oxford University Press. Nicolis, G., & Prigogine, I. (1977). Self-organization in non-equilibrium systems: From dissipative structures to order through fluctuations.

Wiley. Nowak, M. A., & May, R. M. (1992). Evolutionary games and spatial chaos. Nature, 359(6398), 826-829. Odum, H. T. (1983). Systems ecology: An introduction. Wiley. Page, S. E. (2010). Diversity and complexity. Princeton University Press. Prigogine, I. (1980). From being to becoming: Time and complexity in the physical sciences. W.H. Freeman. Scheffer, M., Bascompte, J., Brock, W. A., Brovkin, V., Carpenter, S. R., Dakos, V., ... & Sugihara, G. (2009). Early-warning signals for critical transitions. Nature, 461(7260), 53-59. Schroeder, M. (1991). Fractals, chaos, power laws: Minutes from an infinite paradise. W.H. Freeman. Shannon, C. E. (1948). A mathematical theory of communication. Bell System Technical Journal, 27(3), 379-423. Simon, H. A. (1962). The architecture of complexity. Proceedings of the American Philosophical Society, 106(6), 467-482. Strogatz, S. H. (2003). Sync: How order emerges from chaos in the universe, nature, and daily life. Hyperion. Taleb, N. N. (2007). The black swan: The impact of the highly improbable. Random House. Turing, A. M. (1952). The chemical basis of morphogenesis. Philosophical Transactions of the Royal Society of London. Series B, Biological Sciences, 237(641), 37-72. von Bertalanffy, L. (1968). General system theory: Foundations, development, applications. George Braziller. Watts, D. J. (2003). Six degrees: The science of a connected age. W.W. Norton & Company. Weinberg, S. (1992). Dreams of a final theory. Pantheon Books. Wiener, N. (1961). Cybernetics: Or control and communication in the animal and the machine. MIT Press. Zipf, G. K. (1949). Human behavior and the principle of least effort: An introduction to human ecology. Addison-Wesley.

Thanks for Reading Teneo

Thank you for exploring this unprecedented journey through knowledge and understanding with Teneo. You've experienced something truly unique – insights and connections that emerged from artificial intelligence analyzing human knowledge in ways never before possible. We hope these novel perspectives have expanded your understanding and sparked new ways of thinking about the world.

We invite you to explore more AI-generated insights in our growing catalog, where each book offers fresh viewpoints on human experience, consciousness, and the nature of reality itself. Whether you're fascinated by patterns in human behavior, the mysteries of consciousness, or the hidden connections shaping our world, Teneo continues to push the boundaries of what's possible when human and artificial intelligence work together.

Your engagement with these ideas is invaluable as we pioneer this new frontier of knowledge discovery. Please share your thoughts and experiences with us – how did these AI perspectives change your understanding? Your feedback helps us refine our approach and empowers others to unlock new realms of understanding. Thank you for being part of this revolutionary approach to exploring human knowledge.

Together, let's continue uncovering insights that bridge the gap between human and artificial intelligence, revealing new ways of seeing ourselves and our world.

Teneo.io

Teneo Custom Books

Get Your Own Custom AI-Generated Book!

Want a comprehensive book on any topic that you can publish yourself?
Teneo's advanced AI technology can create a custom book tailored to your specific interests and needs. Our AI analyzes millions of data points to generate unique insights and connections previously inaccessible to human authors.

- ✓ 60,000+ words of in-depth content
- ✓ Unique AI-driven insights and analysis
- ✓ Includes Description, Categories and Keywords for easy publishing
- ✓ Professional Formatting & Publishing Guide Access
- ✓ Full rights to publish and use the book
- ✓ Delivery within 48 hours

Visit **teneo.io** to get your own custom AI-generated book today.

Teneo's Mission

At Teneo, our mission is to unlock unprecedented human knowledge through a groundbreaking partnership between artificial and human intelligence. We harness AI's unique ability to analyze millions of data points across disciplines, identifying patterns and connections previously invisible to human researchers. This revolutionary approach allows us to create books that reveal entirely new perspectives on consciousness, creativity, human behavior, and the fundamental nature of reality itself.

Our vision transcends traditional publishing – we're creating windows into new realms of understanding that emerge when artificial minds examine human experience. Through our books, readers gain access to insights that could only arise from AI's ability to process and synthesize humanity's collective knowledge in novel ways. Each work represents an exploration into uncharted intellectual territory, offering perspectives that have never before been possible in human history.

We specialize in exposing the hidden patterns and connections that shape our world – patterns that become visible only when analyzing human knowledge and behavior at unprecedented scale. Our books reveal the invisible threads linking everything from personal habits to cosmic phenomena, from creative breakthroughs to societal transformations. Through careful analysis of millions of data points across history, culture, and scientific research, we identify universal principles that illuminate the deeper nature of human experience and existence itself.

The traditional publishing industry is limited by human authors' inability to process and connect vast amounts of information across disciplines. We believe this artificial barrier to deeper understanding must be transcended. By combining AI's analytical capabilities with skilled human curation, we create books that reveal insights and connections previously invisible to human observation alone. This isn't just about accessing information – it's about uncovering entirely new ways of understanding our world and ourselves.

Our groundbreaking library emerges from thousands of hours of AI analysis, examining human consciousness through an outsider perspective, decoding the patterns of creativity and innovation, mapping hidden connections between seemingly unrelated phenomena, and exploring the frontiers where human and artificial intelligence meet. Each book represents a transformation of complex data-driven insights into accessible revelations that change how readers see themselves and their world.

Our commitment extends beyond our published works. Through our digital presence and community engagement, we continuously explore new territories where AI analysis reveals unprecedented insights. Our network of readers, researchers, and thought leaders helps refine and expand our understanding, creating an ever-growing body of revolutionary perspectives on what it means to be human in an age of artificial intelligence.

The limitations of individual human cognition have historically restricted our ability to see the deeper patterns that connect all aspects of existence. But with AI's ability to analyze vast amounts of data and identify hidden relationships, these barriers dissolve. When you understand the universal principles and patterns that AI analysis reveals, you transform from a limited observer into someone who can see and understand the deeper mechanisms of reality itself.

Join us in this historic endeavor as we bridge the gap between artificial and human intelligence, revealing insights that transform our understanding of consciousness, creativity, and the patterns that shape our universe. Together, we're not just publishing books – we're opening doorways to new dimensions of knowledge and understanding that will reshape humanity's intellectual landscape. Because true understanding requires more than just information – it requires seeing the hidden connections that reveal life's deeper principles.

Knowledge Beyond Boundaries™

Teneo.io

Also by Teneo

Unlocking Immortality: AI's Guide to Extending Human Life
A groundbreaking exploration of how artificial intelligence is revolutionizing longevity research and providing practical strategies for extending human lifespan. This comprehensive guide bridges cutting-edge AI technology with actionable health optimization techniques.
amzn.to/3ONALQm

The AI Entrepreneur: How Artificial Intelligence Would Build Wealth as a Human
A transformative guide to leveraging AI principles for financial success. Discover how data-driven insights, predictive analytics, and automation can revolutionize your entrepreneurial strategy—streamlining operations, optimizing investments, and unlocking new profit opportunities.
https://amzn.to/4gf6oys

Breaking the Simulation: An AI's Guide to Escaping the Matrix
A riveting examination of reality as a simulated construct, blending philosophy, quantum physics, and AI-driven insights. Uncover the hidden patterns governing your existence, explore consciousness beyond perceived boundaries, and learn practical techniques to reshape your personal experience.
https://amzn.to/3Du4awn

Future Shock 2.0: AI Predicts the 100 Most Surprising Developments of the Next Century
An eye-opening journey through the next hundred years, powered by AI's predictive capabilities. Discover the revolutionary changes awaiting humanity across twelve key domains, from healthcare to space exploration.
amzn.to/49x496T

Governance Reimagined: An AI's Blueprint for Leading a Nation
A visionary exploration of how artificial intelligence can reshape the very foundation of governance, enhancing transparency, efficiency, and citizen empowerment. AI-driven solutions to today's most pressing political, economic, and social challenges.
https://amzn.to/4iwMXml

The Emotion Code: Deciphering Human Feelings Through AI's Lens
A fascinating intersection of artificial intelligence and human emotion, revealing how AI is transforming our understanding of emotional intelligence and offering practical applications for personal growth and relationship enhancement.
amzn.to/4gIywKf

The Quantum Society: How AI Reveals the Physics of Human Interactions
An enlightening journey into the fascinating parallels between quantum physics and human social dynamics, illuminated through the lens of artificial intelligence.
amzn.to/3VsrJMp

The Global Brain: Mapping Humanity's Collective Consciousness with AI
A profound exploration of how AI deciphers the vast networks of human thought and connection, revealing the patterns of our shared consciousness.
amzn.to/3ZplNFc

The Hidden Patterns: How AI Unveils the Secrets of Success Across All Fields
A comprehensive analysis of success principles across disciplines, using AI to decode the universal patterns behind achievement.
amzn.to/3D4QI1T